AF784812

HORTICULTURE: NEW TECHNOLOGIES AND APPLICATIONS

Horticulture: New Technologies and Applications

Editor:

Preethi Kartan

Delve Publishing
www.delvepublishing.com

Horticulture: New Technologies and Applications
Editor: **Preethi Kartan**

© 2017 Delve Publishing LLC

708 3rd Avenue, 6th Floor

New York, NY 10017

United States of America

www.delvepublishing.com

ISBN: 978-1-68095-756-3

Library of Congress Control Number: 2016951512

This book contains information obtained from highly regarded resources. Reprinted material sources are indicated. Copyright for individual articles remains with the authors as indicated. A Wide variety of references are listed. Reasonable efforts have been made to publish reliable data and views articulated in the chapters are those of the individual contributors, and not necessarily those of the editors or publishers. Editors or publishers are not responsible for the accuracy of the information in the published chapters or consequences of their use. The publisher believes no responsibility for any damage or grievance to the persons or property arising out of the use of any materials, instructions, methods or thoughts in the book. The editors and the publisher have attempted to trace the copyright holders of all material reproduced in this publication and apologize to copyright holders if permission has not been obtained. If any copyright holder has not been acknowledged, please write to us so we may rectify.

Notice: Registered trademark of products or corporate names are used only for explanation and identification without intent of infringement.

Delve Publishing LLC publishes wide variety of books and eBooks. For more information about Delve Publishing and its products, visit our website at **www.delvepublishing.com**

About the Editor

Preethi Kartan

Preethi is a postgraduate in Biotechnology from University of Leeds, UK. She is currently working as a Scientific Associate at one of the CRO's in India and her interest lie in life sciences related writing.

Contents

List of Contributors

Rehman HU
Department of Fruit Science, PAU, Ludhiana-141004, India

Gill MIS
Department of Fruit Science, PAU, Ludhiana-141004, India

Melaku Tesfa
Ethiopian Sugar Corporation Research and Training Division, Biotechnology Research Team, Wonji Research Center Wonji, Ethiopia

Belayneh Admassu
National Agricultural Biotechnology Research Program Holetta Agricultural Research Center

Kassahun Bantte
Jimma University College of Agriculture and Veterinary Medicine, Jimma, Ethiopia

Hayfa Jabnoun-Khiareddine
UR13AGR09- Integrated Horticultural Production in the Tunisian Centre- East, Regional Center of Research on Horticulture and Organic Agriculture, University of Sousse, 4042, Chott-Mariem, Tunisia

Rania Aydi Ben Abdallah
UR13AGR09- Integrated Horticultural Production in the Tunisian Centre- East, Regional Center of Research on Horticulture and Organic Agriculture, University of Sousse, 4042, Chott-Mariem, Tunisia
National Agronomic Institute of Tunisia, 1082 Tunis Mahrajène, University of Carthage, Tunisia

Ayed Fakher
Technical Center of Organic Agriculture, 4042, Chott-Mariem, Tunisia

Mouna Gueddes-Chahed
UR13AGR09- Integrated Horticultural Production in the Tunisian Centre- East, Regional Center of Research on Horticulture and Organic Agriculture, University of Sousse, 4042, Chott-Mariem, Tunisia

Ahmed Hajlaoui
UR13AGR09- Integrated Horticultural Production in the Tunisian Centre- East, Regional Center of Research on Horticulture and Organic Agriculture, University of Sousse, 4042, Chott-Mariem, Tunisia

Samir Ben Salem
UR13AGR09- Integrated Horticultural Production in the Tunisian Centre- East, Regional Center of Research on Horticulture and Organic Agriculture, University of Sousse, 4042, Chott-Mariem, Tunisia

Wissem Ben Dhia
UR13AGR09- Integrated Horticultural Production in the Tunisian Centre- East, Regional Center of Research on Horticulture and Organic Agriculture, University of Sousse, 4042, Chott-Mariem, Tunisia

Mejda Daami-Remadi
UR13AGR09- Integrated Horticultural Production in the Tunisian Centre- East, Regional Center of Research on Horticulture and Organic Agriculture, University of Sousse, 4042, Chott-Mariem, Tunisia

Kisan B, Shruthi H
Department of Biotechnology, College of Agriculture, University of Agricultural Sciences, Raichur-584104, India

Sharanagouda H
Department of Biotechnology, College of Agriculture, University of Agricultural Sciences, Raichur-584104, India

Revanappa SB
Department of Biotechnology, College of Agriculture, University of Agricultural Sciences, Raichur-584104, India

Pramod NK
Department of Biotechnology, College of Agriculture, University of Agricultural Sciences, Raichur-584104, India

Elham Hassanpour
Faculty of Agriculture, Horticultural Department, University of Guilan, Rasht, Iran

Jamal-Ali Olfati
Faculty of Agriculture, Horticultural Department, University of Guilan, Rasht, Iran

Mohammad Naqashzadegan
Faculty of Engineering, Mechanic Department, University of Guilan, Rasht, Iran

El-Sharony TF
Horticultural Crops Technology Department, National Research Centre, Dokki, Giza, Egypt

El-Gioushy SF
Department of Horticulture, Faculty of Agriculture, Benha University, Moshtohor, 13736 Qaluiobia, Egypt

Amin OA
Horticultural Crops Technology Department, National Research Centre, Dokki, Giza, Egypt

MA Elhaak
Faculty of Science, Tanta University, Egypt

MZ Matter
Faculty of Science, Minufiya University, Egypt

MA Zayed
Faculty of Science, Minufiya University, Egypt

DA Gad
Faculty of Science, Minufiya University, Egypt

Nolen S
Agricultural Research Station, Virginia State University, Petersburg, Virginia, USA

Zhang B
Virginia Polytechnic Institute and State University, Blacksburg, USA

Kering MK
Agricultural Research Station, Virginia State University, Petersburg, Virginia, USA

Tekle Edossa Kubure
Department of Plant Sciences, College of Agriculture and Veterinary Sciences, PO Box 19, Ambo University, Ambo, West Shoa Zone, Ethiopia

Cherukuri V Raghavaiah
Department of Plant Sciences, College of Agriculture and Veterinary Sciences, PO Box 19, Ambo University, Ambo, West Shoa Zone, Ethiopia

Ibrahim Hamza
Department of Plant Sciences, College of Agriculture and Veterinary Sciences, PO Box 19, Ambo University, Ambo, West Shoa Zone, Ethiopia

Rivera JD
Mississippi Ag and Forestry Exp. Station.-White Sand Unit, Poplarville, MS, USA

Lemus RW
Mississippi State Univ. Dept. of Plant and Soil Sciences, Starkville MS, USA

Gipson ML
Mississippi Ag and Forestry Exp. Station.-White Sand Unit, Poplarville, MS, USA

Gipson RG
Mississippi Ag and Forestry Exp. Station.-White Sand Unit, Poplarville, MS, USA

Shrawan Singh
Vegetable Science, Division of Horticulture and Forestry, Central Agricultural Research Institute (CARI), Port Blair- 744101, Andaman and Nicobar Islands, India

Naresh Kumar Bainsla
Plant Breeding, Division of Field Crops, Central Agricultural Research Institute (CARI), Port Blair- 744101, Andaman and Nicobar Islands, India

Rediet Girma
School of Biosystems and Environmental Engineering, Hawassa University, Ethiopia

Awdenegest Moges
School of Biosystems and Environmental Engineering, Hawassa University, Ethiopia

Shoeb Quraishi
School of Natural Resource and Environmental Engineering, Haramaya University, Ethiopia

Sameh AM Abd-allah
Horticulture Research Institute, a.r.c., Egypt

Peter Hanson
AVRDC—The World Vegetable Center, Shanhua, Tainan 74199, Taiwan

Shu-Fen Lu
AVRDC—The World Vegetable Center, Shanhua, Tainan 74199, Taiwan

Jaw-Fen Wang
AVRDC—The World Vegetable Center, Shanhua, Tainan 74199, Taiwan

Wallace Chen
AVRDC—The World Vegetable Center, Shanhua, Tainan 74199, Taiwan

Lawrence Kenyon
AVRDC—The World Vegetable Center, Shanhua, Tainan 74199, Taiwan

Chee-Wee Tan
AVRDC—The World Vegetable Center, Shanhua, Tainan 74199, Taiwan

Kwee Lian Tee
Universiti Malaysia Sabah, School of Sustainable Agriculture, Sungai Batang Mile 10, 90000 Sandakan, Sabah, Malaysia

Ya-Ying Wang
AVRDC—The World Vegetable Center, Shanhua, Tainan 74199, Taiwan

Yun-Che Hsu
AVRDC—The World Vegetable Center, Shanhua, Tainan 74199, Taiwan

Roland Schafleitner
AVRDC—The World Vegetable Center, Shanhua, Tainan 74199, Taiwan

Dolores Ledesma
AVRDC—The World Vegetable Center, Shanhua, Tainan 74199, Taiwan

Ray-Yu Yang
AVRDC—The World Vegetable Center, Shanhua, Tainan 74199, Taiwan

Adewumi, I.O
Department of Agricultural Engineering, Federal College of Agriculture, I.A.R & T P.M.B 5029 Moor Plantation Ibadan, Oyo-State, Nigeria

Oluwatoyinbo, F.I
Department of Agriculture, Federal College of Agriculture, I.A.R & T P.M.B 5029 Moor Plantation Ibadan, Oyo-State, Nigeria

Ajisegiri, G.O
Department of Industrial and Production Engineering, Faculty of Technology, University of Ibadan, Nigeria

Akinsete, A.E
Department of Electrical and Electronics, Faculty of Technology, University of Ibadan, Nigeria

Preface

Horticulture is defined as the branch of agriculture that deals with the art, science, technology, and business of growing plants. Horticulture involves in production of medicinal plants, fruits, vegetables, nuts, seeds, herbs, sprouts, mushrooms, algae, flowers, seaweeds and non-food crops such as grass and ornamental trees and plants. Plant conservation, landscape restoration, garden design, construction, and maintenance, and arboriculture are part of horticulture. Horticulture is important because of its benefits like high export value, high per unit area yield, high returns per unit area, best utilization of waste land, provision of raw material for industries and protection of environment. It also improves economic conditions of many farmers and has become means of improving livelihood of many under-privileged classes by generating employment. Recent advances in technologies such as biotechnology, genomics, various methods and techniques in cultivation, has accelerated cultivar development greatly in all plants. This book gives insight on latest technologies and their applications in the field of horticulture.

Editor

Preethi Kartan

1
CHAPTER

MICROGRAFTING OF FRUIT CROPS-A REVIEW

Rehman HU and Gill MIS

Department of Fruit Science, PAU, Ludhiana-141004, India

ABSTRACT

Micrografting was developed in 1980s and consists of the placement in aseptic conditions of a maintained scion onto an in vitro grown rootstock. The results of in vitro micrografting and the plant material derived from it can be further multiplied in tissue culture conditions or acclimatized to outdoor conditions. Micrografting is a technique that potentially can combine the advantages of rapid in vitro multiplication with increased productivity that results from grafting, superior rootstock and scion combinations. Among various methods of micrografting, slit or wedge grafting has been found most suitable in case of fruit crops. In vitro shoot tips are better as compared to in vivo shoot tips for carrying out micrografting resulting in higher graft success, less contamination, less vitrification, lower shoot tip necrosis and good vigour of micrografts. Media used for growing micrografted plantlets also affects grafting success, vitrification, contamination and vigour. Use of appropriate medium depends on genotype and growing conditions. Modification of basal medium formulation has been an effective means for achieving good graft success. The sucrose concentration of the nutrient medium of grafted plants has been found to play a significant role. Besides these, scion length and rootstock development stage also influences graft success, vitrification, shoot tip necrosis, vigour and contamination. The use of pre-treated shoot apices not only markedly increases the grafting success but also overcome problems encountered during the handling of in vitro micrografting that results in browning and drying of apices.

INTRODUCTION

The necessity to modernize the planting material production technologies of fruit crops has been stimulated by many considerations such as the trends towards increasing the

Citation: Rehman HU and Gill MIS, Micrografting of Fruit Crops-A Review, doi:10.4172/2376-0354.1000151.

planting densities in field grown trees and the transition to intensive growing systems, which includes the selection of new parents for breeding programmes, the creation and introduction of new cultivars and the modernization of tree habit and pruning. All of these changes have created a demand for more and more quantities of quality planting material. The conventional system of propagating this material is not only time consuming but the material raised is neither uniform nor healthy. Application of *in vitro* techniques in fruit growing can, therefore, be a viable alternative to circumvent these problems. One such application is shoot tip grafting or micrografting. Micrografting was developed in 1980s and consists of the placement of meristem tip or shoots tip explant onto a decapitated rootstock that has been grown aseptically from seed or micropropagated [1]. The results of *in vitro* micrografting and the plant material derived from it can be further multiplied in tissue culture conditions or acclimatized to outdoor conditions. Micrografting is a technique that potentially can combine the advantages of rapid *in vitro* multiplication with increased productivity that results from grafting, superior rootstock and scion combinations [2]. The juvenile phase can be circumvented by micrografting to a rootstock that induces early maturity. Besides to the benefits of traditional grafting, micrografting shoot tips can be an efficient means of regenerating plant material free of endogenous contaminants [3] with enhanced potential for true to type cloning mature plants [4]. Micrografting technique appears to be useful in early screening of graft compatibility in fruit crops as the overlapping of different stress response induced by the graft itself at the interphase between the partners makes identification of the mechanism underlying localized incompatibility difficult [5]. Micrografting has application for physiological analysis of the rejuvenation of mature phase plants [6]. It is possible to graft at any time of the year and to transplant the grafted plants when desired [7].

Grafting under *in vitro* conditions has several advantages both for production and research. *In vitro* shoot tip grafting has often been applied for the (i) improvement and rejuvenation of several tree species [8,9] (ii) virus elimination [10,11] (iii) study of physiological connections between rootstocks and scions such as (in)compatibility, root to shoot communication or transport [12,13] and (iv) use in quarantine as this method has a minimum risk for importing plants [14]. Due to the multiple uses and advantages of shoot tip grafting, this technology may be of interest or of potential practical value to technicians, researchers and nursery operators. Professionals in each of these areas would benefit from the introduction of a simpler and more efficient micrografting procedure that is less dependent on mastery of complex techniques and thus will contribute to the practical utility of micrografting as a tool in fruit tree biotechnology.

REVIEW OF LITERATURE

The literature relevant to micrografting in fruit crops is reviewed under the following headings.

Scion Origin

The origin of the scion can greatly affect the micrografting success. Deogratias et al. [15] studied *in vitro* micrografting of apricots by using actively growing or dormant shoot tips (0.5-1.0 mm) from different sources: (a) dormant buds collected from November to February (b) vegetative flushes from plants growing in the field (c) *in vitro* derived vegetative shoots. The effects of rootstock cultivar, position of the shoot tip on the rootstock, shoot tip size, light and temperature, composition of the medium and growth regulator treatments on the success of grafting on rootstock seedlings growing *in vitro* were studied. The best source of shoot tips was found to be *in vitro* derived micro shoots while, dormant buds collected in November proved inferior one. Many of the grafts became quiescent after initial growth.

Hassanen [16] also reported the highest grafting success while using *in vitro* scion source and obtained grafting success of 83 per cent by using *in vitro* scion (>0.5 cm but <1 cm) of Le Conte pear on *in vitro* decapitated *P betulaefolia* as rootstock.

Bhatt [17] also reported that the best source of shoot tips was from *in vitro* derived shoots and the poorest source was from *in vivo* forced shoot tips in apple. Higher contaminated cultures (39.04%), browning/necrotic cultures (57.15%) and browning/necrosis intensity (3.38) was observed when scions derived from *in vivo* forced tips were used for grafting which decreased to 10.80 per cent, 39.06 per cent and 2.20, respectively when *in vitro* derived scion tips were used.

Dziedzic and Malodobry [18] carried out micrografting using cherry cultivar Regina on rootstocks Damil and Gisela 5 and cv. Van on rootstock Damil. Shoot tip explants of Regina and Van were 0.2-0.3 and 0.4-0.5 cm in length, respectively. Different pre-treatments of scions and stocks, ways of grafting and light conditions after grafting influenced the graft success. Before grafting, the upper part of the rootstocks and the lower end of scions were dipped for one minute in different antioxidant and growth regulator combinations. There was no browning of scion and stocks cut surface following these treatment combinations. Treatment combination of 150 mgl^{-1} citric acid+0.1 mgl^{-1} GA3 +0.5 mgl^{-1} IBA resulted in higher graft success.

Baydar and Celik [19] used *in vitro* micrografting technique in grapes. Shoot tip meristems of KalecikKarasi, Emir, Uslu, Hafizali and Razaki grape cultivars were micrografted onto Kober 5BB seedlings. Shoot tip meristems were obtained from *in vivo* (shoots from actively growing vines in the vineyard or shoots from one-year-old cuttings in the greenhouse) or *in vitro* (shoots obtained by shoot tip culture in aseptic conditions) sources. The micrografted plantlets were cultured on solid MS medium for 2 months. The micrografting success rate was highest using *in vitro* shoot tips, varying from 40.6 per cent in cv. Hafizali to 68.3 per cent in cv. Emir.

Thimmappaiah et al. [20] performed *in vitro* grafting of almond cv. Achak on aseptic rootstocks in order to estimate the success of micrografting in relation to the mode of grafting, the scion origin and the physiological state of the grafts. Both active and dormant small shoots, buds and meristematic apices were collected from field plants. In this study, 4-6 week old shoots and small apical buds (3.0-5.0 mm) formed *in vitro* were also used. These shoots and buds grafted on top or in T-grafts gave a higher percentage (82.1% and 79.2%) respectively of successful grafts. Contamination occurred more often in material taken from the field and necrosis was observed frequently when physiologically active meristematic apices were used. Similarly, Onay et al. [21] reported that the best growth of microscion in Pistachio was obtained with the *in vitro* shoot tips rather than with shoot tips excised from a field grown tree.

Sanjaya et al. [22] were successful in achieving *in vitro* micrografts of Sandal wood (*Santalum album L.*) by placing 1-2 cm long scions derived from nodal segments onto the hypocotyls of 45 days old *in vitro* rootstocks. Use of *in vitro* grown shoots as a source of scion gave better graft success (60%) than scions collected directly from field grown trees. *In vitro* grafting was also influenced by scion size and root age.

Scion Length

Hassanen [16] reported successful micrografting of pear cv. Le Conte on *in vitro* decapitated P betulaefolia as rootstock. Shoot tips consisted of two different sizes, <0.5 cm long and >0.5 cm but <1 cm long were examined. The highest percentage of successful grafts, axillary shoots development percentage, micrograft length increase and number of new buds formation (83%, 80%, 3.5 cm and 4 cm, respectively) were obtained with scion length greater than 0.5 cm, while the scion of smaller than 0.5 cm was achieved 37% of successful grafting.

Rehman and Gill [23] also reported maximum graft success using 5-10 mm scions as compared to <5 mm and 10-15 mm in case of Patharnakh pear when grafted on Kainth rootstock. Besides, graft success per cent, they also reported that scion length have significant influence on necrosis and vigour of micrografts.

It is necessary to choose a shoot tip of size that gives a realistic degree of grafting success. Onay et al. [21] carried out *in vitro* grafting in Pistachio and studied different variables including different size of microscion, grafting methods, effect of culture medium and effect of time of the year at which shoot tips were used. The results indicate that the easiest and most successful method for grafting was slit micrografting. Higher levels of micrograft take were achieved with 2-4 mm (56.76%) and 4-6 mm (79.25%) long scions obtained from the regenerated shoot tips.

Bhatt [17], while working on micrografting of Lal Ambri apple on M-9 rootstock, observed that graft success increased with the size of scions up to 10 mm, then decreased with increase in size. Graft success of 12.92 per cent was obtained when scion tips of 10 mm were used for grafting which decreased to 10.53 per cent and 8.94 per cent when scion tips of 5 mm and 15 mm were used, respectively.

Amiri [24] studied the effect of grafting method, scion size and its origin on grafting success. Grafting success was significantly dependent on the method of grafting and size of the scion. One month old *in vitro* cultured meristematic apices with length of 5-15 mm were used as microscion. The highest percentage (65%) of successful grafts was obtained for homoplastic apex graft (shoot tip), with apical bud scion length greater than 6 mm. Whereas, the lowest percentage (16%) of success was observed in the heteroplastic side bud apices (wedge) grafts. As the size of the microscion increased (from 3 to 15 mm), the success rate of the grafts improved and reached 34 to 65 per cent. Khalafalla and Daffalla [25] noticed that rate of successfully grafted shoots in Gum Arabic tree was influenced by both scion length and rootstock age. Success rate was 100 per cent with scion length of 3.0 cm and rootstock of 14 days.

Thimmappaiah et al. [20] observed that length of scion shoot had significant effect on micrografting success. Graft success was higher (79.5%) when the scion length was greater than 5 mm and it was lesser (0.5%) when size of scion was smaller (3-5 mm) in cashew.

Rootstock Age

Deogratias et al. [26] carried investigations to recover virus free Prunus plants by *in vitro* shoot tip grafting and reported that rootstock age is involved in the success of micrografting. The optimal period to produce meristem resumption is between 7 to 10 days. Earlier than 7 days, rootstock callus invaded the meristem tips and after 10 days, rapid desiccation of rootstock section and scion necrosis occurred.

Dong et al. [27] carried experiments to examine the relationship between the physiological state of the rootstock/scion and grafting success. He revealed that fast-growing tender shoots of walnut had high levels of cytokinin, auxin, glucose and laevulose and formed strong graft unions than older and longer shoots.

Sanjaya et al. [22] carried micrografting of *Santalum album* using *in vitro* and *in vivo* methods that consisted of grafting under aseptic conditions of the miniaturized scion onto a grown rootstock; was conducted and compared the success of the two methods. The frequency of success in shoot apex grafting of *Santalum album* was highest with rootstock and shoots derived from *in vitro* method. The probability of successful graft unions of both *in vitro* and *in vivo* methods increased with increase in length of scion material. However, the success of grafting declined as the age of the seedlings increased, since it was observed that the hypocotyls region became narrower and harder with age. Khafalla and Daffalla [25] observed that the rate of successfully grafted shoots in gum Arabic tree (*Acacia senegal*) was influenced by both scion length and rootstock age. Fourteen days old rootstock was found to be more efficient than seven day old one.

Media Used

Modification of basal medium formulation has been an effective means for achieving graft success. The sucrose concentration of the nutrient medium of grafted plants has been found to play a significant role.

Hosoi et al. [28] reported that sucrose concentration in the liquid medium significantly affected the percentage success in citrus and best results were obtained with 45 gl^{-1} sucrose. Gebhardt and Goldbach [2] described micrografting technique of Prunus shoots derived by shoot tip culture. Autografts of plum cv. Hauszwetsche as well as heterografts of several sour cherry cultivars (Schattenmorelle, Weiroot 158 and Koroser) were established. A mechanically strong graft union was formed during the course of a three week subculture of micrografts in a liquid medium without the addition of growth regulators. Deogratias et al. [15] obtained best results in micrografting of apricot when cultured in medium comprising MS salts with 3 per cent sucrose.

Bhatt et al. [29] while investigating effect of different media formulations on graft success reported highest micrografting success rate using MS semisolid media followed by MS liquid media and MS liquid plus vermiculite media. Higher graft success (24.55% and 21.89%) was obtained when micrografts were cultured on MS semi-solid media at 3 and 6 per cent sucrose, respectively.

Rehman and Gill [23] reported that medium composition and its phase has significant effect on graft success, necrosis, vigour and vitrification of Patharnakh micrografts. Maximum graft success was observed on Murashige and Skoog liquid medium containing 20 g/l sucrose.

Abousalim and Mantell [30] carried out study on micrografting in Pistachio cv. Mateur, on the rootstock of the same cultivar. The scion bases of cv. Mateur were cut into gently sloping wedges and were inserted into vertical splits in the cut surface of the epicotyls. Micrografted seedlings were cultured on liquid MS medium using a filter paper bridge. In second experiment, surface disinfected seeds were cultured directly on top of a Milcap-synthetic (polypropylene fibre) support and after 12 days, the seedlings were micrografted in situ with scions. The best results were obtained using the later method and vascular connections were established across grafts within 21 days. Axillary shoots were produced by 60 per cent of the scions. The Milcap -polypropylene supports allowed good growth and branching of root system and prevented root damage during grafting.

Grewal et al. [31] successfully cultured shoot tip grafted plants of various cultivars of citrus in a liquid medium containing the salts of MS medium, the vitamins of White's medium and sucrose (7.5%). Jonard [6] micrografted several grape cultivars onto Kober 5BB seedlings. These micrografted plantlets were cultured on Murashige and Skoog (MS) medium for two months. Banerjee et al. [32] developed an efficient and highly reproducible *in vitro* micrografting protocol for microshoots of cotton (*Gossypiumhirsutum L.*). Culture of grafted shoots in Murashige and Skoog's basal liquid medium devoid of growth regulators showed maximum rate of survival (95%), elongation of scions and number of nodes per plantThimmappaiah et al. [20] developed a successful micrografting technique in cashew using *in vitro* germinated seedlings as rootstocks and axenic shoot cultures (shoot tip and nodal cultures) established from mature tree as microscions. *In vitro* germinated seedlings which emerged 20-25 days after inoculation on absorbent cotton were decapitated and used as rootstock. Mature tree explants initiated on hormone-free [33] modified medium were prepared as scions of 3-15 mm length for grafting. Micrografts were successfully cultured on hormone-free liquid half MS medium and potted out after 10-12 weeks of culture growth. Dziedzic et al. [33] carried out micrografting of cv. Regina on Gisela 5 rootstock on two types of MS media prepared as liquid and semi-solid with 0.7 per cent agar. High per cent of successful micrografts were achieved on semi-solid MS medium. Higher contaminated cultures were observed in liquid media. However, vigorous plantlets developed in this liquid media. Naz et al. [34] carried study to assess the potential use and applicability of micrografting technique for the development of virus free nursery in citrus. Some techniques that tend to increase the grafting success were employed. MS medium fortified with 3, 5 and 7 per cent sugar was used in combination with two grafting methods viz. inverted-T incision and surface placement in Kinnow mandarin and Succari sweet orange. Grafting was carried out under aseptic conditions by using

15 day old etiolated seedlings of rough lemon as rootstock. Shoot tips (1-2 mm) and 3 leaf primordia (0.3-0.5 mm) were taken from the fresh shoot flushes and grafted *in vitro*. Higher grafting success of 34.7 per cent was recorded with inverted-T incision than surface placement which gave 26.7 per cent successful micrografts. A total of 21 per cent successful micrografts were achieved at 3 per cent sugar level which increased significantly to 33 per cent with increase in sugar level to 5 per cent in both cultivars.

Graft Method

The method of placement of the excised shoot tip onto the rootstock significantly influences the grafting success. Various methods of placement of scion tip onto rootstock have been described. Abreu et al. [35] studied micrografting of tissue cultured apple plants. Scions were cleft-grafted on the rootstocks of Maluspumila and Malusprunifolia. Graft development resulted in the generation of meristem tissue with parenchymatous cells originating at the graft interface and cambial tissue proliferation from the scion. This allowed a vascular system connection between the scion and rootstock, resulting in micrograft viability.

Bhatt et al. [17] and Bhatt et al. [29] observed that vertical slit grafting as best method for carrying out micrografting in apple. He reported highest graft success (23.83%) and lowest browning/ necrotic cultures (29.33%) were obtained with vertical slit method. The graft success decreased to 21.00 per cent and 13.10 per cent and browning/necrosis incidence per cent increased to 34.68 per cent and 40.36 per cent when scion was grafted by wedge and horizontal placement, respectively.

Rehman and Gill et al. [23] observed wedge grafting to be better than horizontal grafting in terms of graft success, incidence of necrosis and vitrification. Graft success reached upto 33.71% when Patharnakh scions of 5-10 mm were micrografted using wedge grafting technique allowed to grow in Murashige and Skoog liquid medium. Bhatt et al. [32] reported the highest vigorous plantlets resulted by carrying vertical slit grafting techniques compared to wedge and horizontal grafting.

Onay et al. [21] studied the influence of different grafting methods on graft success in Pistachio. He observed that the slit method was easiest and most successful one giving success of 80 per cent whereas 60 per cent graft success was obtained under wedge micrografting. In vertical slit micrografting no problem was observed during the union formation as long as the contact surfaces were perfectly smooth.

Dobranszki et al. [36] described the requirements for *in vitro* micrografting in apple cv. Royal Gala which served as source of both rootstock and scion. Oxidative browning of cut surface was inhibited by the use of antioxidant mixture during grafting process. Scion base of Royal Gala cut in V-shape was dipped in 1.0 per cent agar-agar solution and stuck into the vertical slit of rootstock. There was no displacement and the rate of fused and further developed grafts were 95 per cent.

Naz et al. [34] investigated the applicability of micrografting technique for the development of virus free nursery in citrus and reported that success in micrografting varies with the method and plant species or genotype. Graft success of 34.7 per cent was recorded with inverted-T incision whereas; surface placement gave 26.7 per cent successful micrografts. Succari sweet orange responded maximum with surface placement method whereas, Kinnow mandarin showed better response with inverted T incision [37].

Toth et al. [38] studied the efficiency of *in vitro* grafting method of six plant species-pepper, melon, apricot, peach, tobacco and chrysanthemum. Plants with short roots were removed from the media and the stem of the scion and the stock (at the four to five leaf stage) were cut at right angles, each with 2-3 leaves remaining on the stem. The stem of the scion was cut in a wedge with scalpel in Petri-dish and the tapered end was fitted into a cleft made in the cut end of stock. The graft was firmed with silicon rubber clip hauled up with dentist's forceps. The survival rate of the grafted plants was excellent in case of herbaceous plants (90-100%) whereas it was 30 per cent for woody species.

Dziedzic and Malodobry [18] carried out micrografting using cherry cv. Van and Regina on Damil rootstock using different methods of grafting. The wedge shaped end of scion 'Van' was inserted into 0.5-0.6 cm slit in rootstock Damil, whereas, even end of shoot tip of Regina was placed on Damil rootstock. The grafts were placed individually into test tubes on WPM medium, with 30 gl^{-1} sucrose and agar 7gl^{-1}. High per cent of successful grafts were obtained for cleft grafting method than shoot tip grafting. The scions and stocks were attached to each other closely and the per cent of graft unit displacement was low.

Wu et al. [39] developed shoot tip micrografting technique in Proteacynaroides wherein, seedlings were decapitated and a vertical incision made from the top end. The axenic microshoots were cut into a wedge (V) shape and placed them into the vertical incision of the rootstock. Best results were obtained by placing the microscions directly onto the rootstocks without any pre-treatment. Hassanen et al. [16] also reported that grafting success was significantly influenced by grafting method employed during micrograting of pear cv. Le Conte on *P betulaefolia*.

VITRIFICATION

Micropropagation of woody plants is often hampered by the phenomenon of hyperhydricity or vitrification [40]. Hyperhydricity, which is characterized by a glassy or swollen appearance to the tissue, usually results in reduced multiplication rates, poor quality shoots and tissue necrosis [41]. The so-called vitrified, vitreous or hyperhydric shoots appear turgid, watery at their surface and hypolignified. Their organs are translucent and in some cases less green and easily breakable.

Occurrence of hyperhydricity remains unpredictable and many factors appear to induce hyperhydricity [42]. Stress is believed to be the major factor underlying the phenomenon. Stress during *in vitro* culture could arise as a result of wounding, the properties of the tissue culture medium, infiltration of tissue by the culture medium which is generally of a higher ionic strength and rich in nitrogen or environmental conditions during *in vitro* culture in sealed containers [42]. This stress could mediate a rapid endogenous ethylene burst. One of the reasons of poor success rate in micrografting is vitrification.

Bhatt [17] reported that media formulations significantly influences vitrification percentage of apple. Lowest vitrified cultures (4.94%) were obtained in MS semi- solid media with 6 per cent sucrose which increased non-significantly to (6.29%) in the same media but with 3 per cent sucrose. Highest vitrified cultures (16.32% and 12.58%) were observed in liquid medium at 3 per cent and 6 per cent, sucrose respectively. MS liquid medium plus vermiculite also showed significantly higher vitrified cultures than obtained in MS semi solid media.

Amiri [24] reported that initial percentage of successful grafts of cherry (Prunusavium L.) var. Seeyane Mashad was about 75 per cent which decreased at the end of experiment. The poor success rate was mainly due to excessive collogenesis observed on more than 50 per cent of the apices and vitrification on more than 30 per cent of the explants.

Al-Maarri and Al-Ghamdi [43] noticed that the frequency of vitrification increased on medium containing a low concentration of sucrose. Ghorbel et al. [37] studied *in vitro* grafting of almond cv. Achak performed on asceptic roots in order to estimate the success of micrografting in relation to the mode of grafting, the scion origin and physiological state of grafts. He achieved low success rates due to vitrification in more than 70 per cent of explants. Dziedzic [33] carried out micrograting experiments in cherry cv. Regina on Gisela-5 rootstock on two types of Murashige and Skoog medium: liquid medium and semi-solid agar medium. High per cent of successful micrografts were obtained on MS semi-solid medium. Use of liquid media resulted in maximum vitrified plantlets.

Pretreatment of growth regulators

The success of micrografting to obtain disease free plant material depends on various factors; among the many other factors pre-treatment with growth regulators have resulted in speeding up the healing of tissue at graft union, resulting in higher survival of grafts. The use of pre-treated apices not only markedly increases the grafting success but also overcome problems encountered during the handling of *in vitro*micrografting that results in browning and drying of apices. Together with auxins, cytokinins stimulate both apex initiation and the union between rootstock and scion by increasing vascular bundle formation [44]. Cytokinins rejuvenate plant cells and stimulate cell proliferation of graft union tissue with auxins [45].

Bhatt [17] reported treatment of LalAmbrimicroscions with combination of Kinetin and 2,4-D at 0.5 mgl^{-1} and 5 mgl^{-1}, respectively before grafting on M-9 showed highest graft success (42.25%) and lowest browning/ necrosis incidence (19.00%) and browning/necrosis intensity (0.20).

Starrantino et al. [46] reported that best results in citrus were obtained when both components of graft were pre-treated with 0.5 ppm BA, 0.5 ppm Kinetin or 10 ppm 2, 4-D. Lemons gave the highest percentage of successful grafts (88-92%) and clementines the lowest (63-73 per cent).

Edriss and Burger [47] studied the various factors affecting the success of shoot tip grafting. Growth regulator pre-treatment of scions with 2, 4-D and Kinetin was accomplished by dipping the shoot apex of 'Mexican' lime, 'Valencia' orange and 'Star Ruby' grape fruit in growth regulator solutions just prior to excision of the apical dome and primordial leaves. Dipping the shoot tip in 2, 4-D (10 mgl^{-1}) or Kinetin (1 mgl^{-1}) before grafting nearly doubled the success rate compared to control. High concentration of Kinetin (10 mgl^{-1}) and/or low concentration of 2, 4-D (1 mgl^{-1}) were no more effective than the control. There were no differences among the 3 rootstocks in their response to pre-treatments.

Jonard et al. [48] reported that inserting an agar block containing different plant growth regulators resulted in 100 per cent successful *in vitro*grafts of Eureka lemon on Troyer Citrange. Apex pre-treatment on [35] medium plus 3 per cent sucrose with GA3 at 2.9 m moles/litre, BA at 44 mmoles/litre and 2, 4-D at 0.009 mmoles/litre.

Starrantino and Caruso [46] carried out micrografting experiments in citrus species with several growth substances and observed that dipping the apex and the decapitated seedlings for 10 minutes before micrograting in a solution of BA (0.5 ppm) increased the percentage rate of sprouting from 73 to 91 per cent.

Parthasarathy et al. [49] conducted experiments to standardize procedures for *in vitro* grafting of Khasi mandarin on various rootstocks. The immersion of Khasi mandarin shoot tips in 2,4-D (10 μgl^{-1}) for five minutes before grafting improved the grafting success on Cleoptara mandarin, Rangpur lime, Kagzi lime and Khashi mandarin. Histological studies using a scanning electron microscope showed callus development between rootstock and scion. Good graft union and complete vascular connection between the shoot tips and the rootstocks were achieved.

Kumarin et al. [50] reported the influence of pre-treatments of growth regulators on the success of graft take in Nagpur mandarin cultivar on rough lemon rootstock. Just prior to grafting, both rootstocks and scions were dipped in growth regulator solutions for 10 minutes. The solutions tested were BA (0.5, 1.0, 2.0 and 5.0 mgl^{-1}), Kinetin (0.5, 1.0, 2.0 and 5.0 mgl^{-1}) and 2,4-D (1.0, 2.0, 5.0 and 10.0 mgl^{-1}). Grafting success was highest with Kinetin (1.0 mgl^{-1}) or 2,4-D (10.0 mgl^{-1}).

Nunes et al. [51] studied the micrografting technique in apple by using *in vitro* plants and verifying the influence of application of growth regulators IBA and BAP in Gala/M-9 and Gala/Marubakaido as scion rootstock combinations. The percentage of viable micrografts for the connection of vascular tissues, varied from 73 to 93.2 per cent. The percentage of viability observed highest in Gala/M-9 (93.2%) on ½ MS medium supplemented with 2.2 mM IBA, followed by ½ MS medium pre-treatment with IBA (4 mM) in both combinations showed (86.4%) viability. Results showed that treatment with growth regulators at the point of grafting significantly increased the viability of the graft.

CONCLUSION

Micrografting is a technique that potentially can combine the advantages of rapid *in vitro* multiplication with increased productivity that results from grafting, superior rootstock and scion combinations. Graft success has been reported to be influenced by large number of factor individually and as well as in interactive manner such as scion length, scion origin, graft method, media used and rootstock development stage. Grafting under *in vitro* conditions has several advantages both for production and research. *In vitro* shoot tip grafting has often been applied for the improvement and rejuvenation of several fruit tree species, virus elimination, study of physiological connections between rootstocks and scions such as (in) compatibility root to shoot communication or transport and use in quarantine as this method has a minimum risk for importing plants. Micrografting has application for physiological analysis of the rejuvenation of mature phase plants. It is possible to graft at any time of the year and to transplant the grafted plants when desired. Due to the multiple uses and advantages of micrografting, this technology may be of interest or of potential practical value to technicians, researchers and nursery operators.

REFERENCES

1. Gebhardt K, GoldbachH (1988) Establishment, graft union characteristics and growth of Prunusmicrografts. PhysioPlant 72: 153-159.
2. Zilka SE, Faingersh E, Rotbaum A, Tam Y, Spiegel S, et al. (2002) In vitro production of virus free pear plants. ActaHort 596: 477-479.
3. Franclet A (1983) Rejuvenation : theory and practical experiences in clonal silviculture. In :Clonal Forests : Its Impact on Tree Improvement and Our Future Forests (Eds. L. Suffa, R.L. Rauter, W.W. Yeatman). Toronto: Canadian Tree Improvement Association, pp 96-134.
4. Pirovano L, Sacchi GA, Aburuzzese A, Cocucci M (2002) Study on gene expression in compatible and incompatible pear/pear and pear/quince grafts by mRNA differential display. ActaHort 596: 393-395.
5. Jonard R (1986) Micrografting and its application to tree improvement. In: Biotechnology in Agriculture and Forestry (Ed) YPS Bajaj. Trees I-Springer-Verlag, Berlin, Heidelberg, New York.
6. Musacchi S, Ganino T, Grani S, Fabbri A (2004) Grafting in vitro shoots on acclimating plant : Morphological and anatomical aspects of compatible and incompatible pear/quince combinations. ActaHort 658: 559-63.
7. Parthasarathy VA, Nagaraju V, Rahman SAS (1997) In vitro grafting of Citrus reticulate Blanco. Folia Hort 9: 87-90.
8. Canon C, Mehmet O, Hakan T, Kamil S, Elman I (2006) In vitro micrografting of pistachio (Pistaciavera L.) var. Siirt on wild pistachio rootstocks. Journal of Cell and Molecular Biology 5: 25-31.
9. Bandeira SF, Xavier A, Otoni WC, Days JMM (2006) In vitro grafting in the propagation of Eucalyptus urophylla x Eucalyptus grandis clones. Agropecuaira research Brazilian41: 223-232.
10. Bisognin C, Ciccotti AM, Moser M, Grando MS, Jarausch W (2008)Establishment of an in vitro screening system of apple proliferation–resistant rootstock genotypes based on micrografting. ActaHort 781: 375-380.

11. Ribeiro LM, Peixoto JR, Andrade SRM, Fonseca RS, Vieira LM, et al. (2008) Ex vitro micrografting aiming the CABMV virus elimination in passion fruit plant. Pesquisa. AgropecuariaBrasileira 43: 589-594
12. Nelson T (2004) Plant signalling: notes from the underground. Current Biology 14: 929-930.
13. Bortolotti C, Murillo L, Fontanet P, Coca M, Segundo BS (2005) Long-distance transport of the maize pathogenesis-related PRms protein through the phloem in transgenic tobacco plants. Plant Science 168: 813-821.
14. George EF (1993) Micrografting. In: George, E.F. (ed.). Plant propagation by tissue culture. Part-1. Exegetics Ltd. EdingtonWilts., UK, pp. 64-66.
15. Deogratias JM, Castellani V, Dosba F, Jaurez, J, Arregui JM, et al. (1991) Study of growth parameters on apricot shoot-tip grafting in vitro (STG). ActaHorticulturae 293: 363-371
16. Hassanen SA (2013) In vitro grafting of pear (Pyrus spp.) World Applied Sciences Journal 21: 705-709.
17. Bhatt KM (2008) In vitro micrografting of apple (MalusdomesticaBorkh.) cv. LalAmbri on dwarfing M-9 rootstock. Ph.D thesis submitted to Sher-e-Kashmir University of Agricultural Sciences and Technology, Shalimar, Srinagar, J&K.
18. Dziedzic E, Malodobry M (2006) Vegetative cherry rootstocks in tissue culture. SodininkysteIrDarzininkyste 25: 77-84.
19. Baydar N G and Celik H (1999) The effect of shoot tip source on the success of in vitro micrografting in grapevine (VitisviniferaL.). Turkish Journal of Agriculture and Forestry 23: 741-747
20. Thimmappaiah, Puthra GT, Anil SR (2002) In vitro grafting of cashew (Anacardiumoccidentale L.) SciHort 92: 177-182.
21. Onay A,PirincV, AdiyamanF, Isikalen C, TilkatEand Basaren D (2004)In vivo and In vitro micrografting of pistachio (Pistaciavera L. cv. 'Siirt'). Turkish J Bio 27: 95-100
22. Sanjaya, Ananthapadmanabho HS, Ravishankar R, Rai R (2003) In-vitro and in-vivo micrografting of (Santalum album L.) shoot tips. Journal of Tropical Forest Science 15: 234-236.
23. Rehman HU and Gill MIS (2014) In vitro shoot tip grafting of Patharnakh [Pyruspyrifolia (Burm F.) Nakai] pear on Kainth rootstock. Vegetos 27: 363-369.
24. Amiri ME (2006) In vitro techniques to study the shoot tip grafting of PrunusaviumL. (cherry) var. SeeyaheMashad. Journal of Food Agriculture and Environment 4: 151-154.
25. Khafalla MM, Dafalla HM (2008) In vitro micropropagation and micrografting of gum Arabic tree (Acacia Senegal L.) wild. InterJ Sustainable Crop Prod 3: 19-27.
26. Deogratias JM, Lutz A and Dosba F (1986) In vitro micrografting of shoot tips from juvenile and adult Prunusavium L. and Prunuspersica L. Batsch to produce virus free plants. ActaHorticulturae 193: 139-145.

27. Dong P, ShengKe X, Dong-F, Pei D, Xi-SK, Dong FX (1998) Studies on some scion-rootstock physiological indexes related to micrografting survival of walnut. Forest Research 11: 119-23.

28. Hosoi T, Machida H, Ooishi A, Miura T (1979) Studies on shoot tip grafting of citrus. Bulletin of the Faculty of Agriculture, Shizuoka University 29: 11-17

29. Bhatt KM, Banday FA, Mir MA, Rather ZA,Hussain G (2013) In vitro grafting in apple(Malusdomestica. Borkh) cv. LalAmbri. Karnataka Journal ofAgricultural Sciences26: 399-402

30. Abousalim A, Mantell SH (1992) Micrografting of pistachio (Pistaciavera L. cv. Mateur). Plant Cell Tissue and Organ Culture 29: 231-34.

31. Grewal HS, Dhatt AS, Gosal SS (1994) In vitro shoot tip grafting in citrus. Annals of Biology, Ludhiana 10: 1-6.

32. Banerjee AK, Agrawal DC, Nalawade SM, Krishnamurthy K V (2000) Recovery of in vitro cotton shoots through micrografting. Current Science 78: 623-26.

33. Dziedzic E (2004) Preliminary studies on vaccination cherry (Prunusavium L.) in vitro. The film UniversitatisAgriculturaeStetinensis, cheese. Agricultura 240: 41-44.

34. Naz AA, Jaskani MJ, Haider A, Qasim M (2007) In vitro studies on micrografting technique in two cultivars of citrus to produce virus free plants. Pakistan Journal of Botany 39: 1773-78.

35. Abreu MF, Nunes JCO, Santos M, Pedrotti EL (2003) Preliminaries histological studies from the micrografting of tissue cultured apple. Revista-Brasileira-de-Fruticultura 25: 195-96.

36. Dobranszki J, Magyar TK, Jambor BE, Lazanyi J (2000) New in vitro micrografting method for apple by sticking. International Journal of Horticultural Sciences 6: 79-83.

37. Ghorbel A, Chatibi A, Thaminy S, Kchouk ML, et al (1998) Micrografting of almond [Prunusdulcis (Miller) D. A. Webb] cv. Achak. ActaHort 470: 429-33.

38. Toth EK, Kriston E, Csillery G, Csanyi ME (2006) Application of silicon rubber grafting clips for in vitro phytopathological studies. ActaHort 725: 373-78

39. Wu HC, Toit ESD, Reinhardt CF (2007) Micrografting of Proteacynaroides. Plant Cell Tissue and Organ Culture 89: 23-28.

40. Chakrabarty D, Park SY, Ali MB, Shin KS, Paek KY (2005) Hyperhydricity in apple: ultra structural and physiological aspects. Tree Physiology 26: 377-88.

41. Ziv M (1991) Vitrification: morphological and physiological disorders of in vitro plants. In: Micropropagation Technology and Applications. Eds. P.C. Debergh and R.H. Zimmerman, Kluwer Academic Publisher, Dordrecht, Netherlands, pp. 71-93.

42. Kevers C, Coumans M, Coumans GMF, Gaspar T(1984) Physiological and biochemical events leading to vitrification of plants cultured in vitro. PhysiologiaPlantarum 61: 69-74.

43. Al-Maarri K , Al-Ghamdi, AS (1996) Factors affecting the incidence of vitrification of in vitro propagated fruit trees. Journal of King Saud. University 8: 139-49.

44. Quoirin M, Lepoivre P (1977) Improved media for in vitro culture of Prunussp. ActaHort 78 : 437-42 (Original not seen. Cited by Bell R L and Reed B M, 2002, ActaHort 596: 412-18

45. Bessis R (1986) Grapevine Physiology: the contribution of culture in vitro. Cellular and Molecular Life Sciences: 927-33

46. Starrantino A and Caruso A (1988)The shoot-tip grafting technique applied in citriculture. ActaHort 227: 101-03.

47. Edriss MH, Burger DW (1984) Micrografting shoot tip culture of citrus on three trifoliate rootstocks. ScientiaHort 23: 255-59.

48. Jonard R, Soedharma I, Villemur P (1987) Analysis of the influence of various factors on the improvement of successful micrografts in citrus. Comptes-Rendus-de-l'Academie-des-Sciences, III Sciences-de-la-Vie 305: 45-49.

49. Parthasarathy VA, Nagaraju V and Rahman SAS(1997)In vitro grafting of Citrus reticulataBlanco. Folia Hort 9: 87-90.

50. Kumarin V, Shyam S, Kumarin V, Singh S (2000) Shoot tip grafting with growth regulators for virus elimination in Nagpur mandarin (Citrus reticulata). Indian Journal of Agricultural Sciences 70: 396-97.

51. Nunes JCO, Abren MF, Dantas ACM, Pereira AJ, Pedrotti EL (2005) Morphologic characterization in apple micrografts. RevistaBarasileira de Fruticultura 27: 80-83.

2

CHAPTER

EX VITRO ROOTING OF SUGARCANE (SACCHARUM OFFICINARUM L.) PLANTLETS DERIVED FROM TISSUE CULTURE

Melaku Tesfa[1] Belayneh Admassu[2] Kassahun Bantte[3]

[1]Ethiopian Sugar Corporation Research and Training Division, Biotechnology Research Team, Wonji Research Center Wonji, Ethiopia

[2].National Agricultural Biotechnology Research Program Holetta Agricultural Research Center

[3].Jimma University College of Agriculture and Veterinary Medicine, Jimma, Ethiopia

ABSTRACT

A study was conducted at Holetta National Agricultural Biotechnology Laboratory with the objective to determine the effect of different concentrations of NAA on ex vitro root development of sugarcane microshoots. Five levels of NAA (0, 10, 20, 30 and 40 mg/l) and two levels of genotypes were combined in factorial arrangement. The basal end of the shoots was dipped in NAA solution overnight before the shoots were transferred into a plastic tray containing a mixed growing medium in green house. The results showed that interaction effect of genotype and NAA was highly significant ($p<0.0001$) on rooting percentage, number of roots per shoot, root length. In genotype N52, best root formation was found on the shoots treated with 20 mg/l NAA by which rooting percentage was 76 ± 5.48 with 5.88 ± 0.04 cm root length and 8.06 ± 0.13 number of roots per plantlets. While in genotype N53 maximum, root formation was recorded on the shoots dipped in 30 mg/l NAA by which rooting percentage was 70 ± 7.07 with 5.42 ± 0.11 cm root length and 4.52 ± 0.19 number of roots per plantlets. Shoots rooted through this method exhibited 100 % survival in both genotypes.

Citation: Melaku Tesfa, Belayneh Admassu Kassahun Bantte, Ex Vitro Rooting of Sugarcane (Saccharum officinarum L.) Plantlets Derived from http://dx.doi.org/10.4172/2329-8863.1000215

INTRODUCTION

Sugarcane (Saccharum officinarum L.) is a monocotyledonous crop that is grown in the tropical and subtropical regions of the world, and almost cultivated on over 23.8 million ha for its sucrose rich stalk [1]. In Ethiopia, it is cultivated commercially on around 96,000 ha with annual sugar production of 370,000 ton and is a solely raw material for sugar production in the country. Recently, in Ethiopia, sugarcane production has gained attention, allied with its important potential for an environment-friendly bio-fuel (ethanol) production, in creating job opportunity to the nation and generating huge electric power [2]. Sugarcane, since commercially propagated vegetatively by stem cutting, has a low seed multiplication rate (1:10) which resulted in slow seed production of newly released improved varieties. Furthermore, the seed builds up diseases and pests during several cycles of field production, which leads to further yield and quality declines over years [3]. Thus, unavailability of disease-free, true to type planting material is a major limitation in improving sugarcane productivity. Recently, tissue culture technology plays a leading role in rapid multiplication of disease-free and quality planting material of sugarcane [4]. Accordingly, Ethiopian Sugar Corporation has established its own tissue culture laboratory at Metahara, Kuraz, Tendaho and Fincha sugar factories/projects to produce about 55 million disease free plantlets per year [2]. To do so, in vitro propagation protocols have been developed using shoot tip and callus explant for several sugarcane cultivars of Ethiopian Sugar Estates [5-8]. In vitro propagation involves four crucial steps namely, initiation, multiplication and rooting of microshoots and acclimatization of plantlets. However, in vitro rooting process is an expensive, labour consuming process and can even double the final price of micropropagated plants. Earlier report shown that in vitro rooting may account for 40% total cost of the intensive manipulation needed during in vitro propagation [9]. In addition, roots of plantlets raised in vitro are generally very fragile and not have root hairs [10]. Therefore, during early acclimatization period, the roots do not function properly to support the plantlets to absorb water and nutrients from the potting medium. Ex vitro rooting is more advantageous than in vitro rooting in reducing cost of labour, chemicals and equipments, and the time of establishment from laboratory to soil revealed that ex vitro rooting reduced more than 50% cost of sugarcane plantlet raised by conventional micropropagation [11-13]. Besides, the plantlets produced after ex vitro rooting have better developed root system than the ones produced after in vitro rooting [14,15]. Furthermore, rooting and acclimatization phase can be carried out simultaneously; hence, it is more time efficient. Ex vitro rooting has been applied in micropropagation of various plants species [11,12,16], but there are very limited reports yet on ex vitro rooting of sugarcane plantlets. Most reports of ex vitro rooting of plant species have involved treatment with exogenous auxin such as indole-3-acetic acid (IAA), indole-3-butyric acid (IBA), and 1-naphthalene-acetic acid (NAA) [12,13,16]. Auxin is applied singly or in a combination at different concentrations to improve rooting frequency of plantlets in the acclimatization period. Therefore, this study was aimed to determine the effect of different concentrations of auxin, NAA on ex vitro rooting of two elite sugarcane genotypes.

MATERIALS AND METHODS

The study was undertaken at the National Agricultural Biotechnology Laboratory of the Ethiopian Institute of Agricultural Research, in Holetta. The study was undertaken with two elite sugarcane genotypes viz. N-52 and N-53 obtained from Ethiopian Sugar Corporation, Research and Training Division. The genotypes were selected based on their higher yield performance and sugar quality. In order to carry out explant sterilization and preparation, actively growing shoot tops were excised from 5-months-old screen house grown healthy mother plants of genotype N52 and N53. The trimmed shoot tops (segment) were washed carefully under running tap water for 30 minutes, and reduced to 10 cm length by cutting off at the two ends. Then washed thoroughly for 30 minutes with tap water containing a drop of liquid detergent solution and two drops of tween-20 and rinsed three times with double distilled water. Subsequently, the explant was taken to a laminar air flow cabinet and immersed in 0.1% (w/v) Bavistin® DF 50 % (Carbendizem) fungicide solution, ascorbic acid (0.2% w/v) and citric acid (0.4% w/v) for 30 minutes followed by three times rinsing each for five minutes with sterile double distilled water. The shoot tips were washed again with 70% ethanol for one minute and rinsed with sterile double distilled water three times each for five minute to remove residual ethanol from the shoot tip surface. Finally, surface sterilized with 50 % (v/v) aqueous solution of Sodium hypochlorite (5.25% w/v active chlorine) containing a few drops of tween-20 for 25 minutes. After pouring out sodium hypochlorite solution, the explants were rinsed with sterile double distilled water three times each for five minutes to remove all the trace of the sterilant. Culture were initiated on [17] medium fortified with BAP, Kinetin and NAA (0.5 mg/l each) [18] and 2% sucrose (w/v) and solidified using agar (agar agar, type I) (0.45 %; w/v) for induction of shoots. The pH of the medium was adjusted to 5.8 followed by autoclaving at 121°C at 105 Kpa pressure for 20 minutes. After 30 days of inoculation the regenerated shoots were transferred to multiplication medium supplemented with 2 mg/l BAP + 0.5 mg/l Kinetin (N52) and 1.5 mg/l BAP and 0.5 mg/l Kinetin (N53) [19]. After 30 days of incubation, shoots were maintained on plant growth regulators (PGRs) free MS medium with 2 g/l activated charcoal for two weeks before transferring rooting stage in order to avoid the carry over effect of hormones from the multiplication media on ex vitro rooting of the plantlets.

Table 1: Effect of different concentration of NAA on ex vitro rooting.

Source of variation	DF	Rooting Percentage	Root length (cm)	Number of roots per shoot
		MS	MS	MS
Genotype	1	968***	7.76***	59.19***
NAA	4	2307***	4.73***	25.56***
Genotype*NAA	4	163***	2.08***	4.46 ***
CV %		10.38	5.60	2.72

After 45 days, healthy micro-shoots having 4 cm heights were employed for ex vitro rooting study. The clumps of in vitro shoots were separated to obtain single micro shoot. The basal portion of these rootless microshoots was dipped in distilled aqueous solution containing auxin, NAA at different concentrations i.e. 0, 10, 20, 30, & 40 mg/l overnight to induce rooting under ex vitro condition. The experiment was arranged in completely randomized design (CRD) with five replications and each treatment had 50 microshoots. After treated with auxins, the shoots were transferred to polystyrene trays containing autoclaved mixture of river sand and forest soil in 2:1 ratio. Subsequently, maintained in greenhouse, which uses Fan-Pad evaporative cooling system providing 25–30° C temperature. During experimenting, high humidity level (80 %-85 %) was maintained by covering the tray with moisten polyethylene sheet and red shade cloth and then sprinkled with water three times a day as necessary and sprayed with quarter strength MS basal medium at weekly interval. After 4 weeks, the plantlets were carefully removed from the soil mix and data on number of rooted shoots, total number of primary roots and root length were recorded. All microshoots that remain green were considered living and used in calculating rooting percentage. Successfully rooted plantlets were subsequently transferred in medium polyethene bags (15 cm × 20 cm) containing mixture of sand, farm yard manure and soil in 1:1:1 ratio for further hardening and data on survival rate of the plantlets was recorded 4 weeks after transplanting. The collected data were subjected to analysis of variance (F test) using SAS program (Version 9.2). The differences among treatment means were determined by REGQ multiple range test at $P<0.05$.

Results and Discussion

Statistical analysis of variance showed that the main effect of genotype and NAA and the interaction effect of genotype and NAA highly significant $(p<0.0001)$ on rooting percentage, number of roots per shoot, root length of the two sugarcane genotype (Table 1). The present result also showed that rooting was induced ex vitro over the entire range of NAA concentration tested including the control shoots in both sugarcane genotypes (Table 2). In the control treatment reduced rooting frequency of 36% and 28% were obtained in genotypes N52 and N53, respectively (Figure 1A and Figure 1B) However, in NAA treated microshoots than 50% of the shoot developed roots regardless of the NAA concentration (Table 2). Shekafandeh [12] observed increased rooting frequency and number of roots from zero percent in untreated shoots to 91.7% and 3.3 roots per shoot, respectively, when the basal end of the shoots were dipped in a solution of 1.5 mg/l IAA and 0.3 mg/l IBA for 24 h before culturing in soil mixture in Myrtle (Myrtus communis L.) plant. Similar results were also reported by Sumaryono and Riyadi [20] in oil palm (Elaeis

Table 1: Effect of different concentration of NAA on ex vitro rooting.

Treatment	Genotypes					
	N52			N53		
NAA mg/l	Rooting Percentage	Root length (cm)	Number of root per shoot	Rooting Percentage	Root length (cm)	Number of root per shoot
0	$36^{d} \pm 5.48$	$4.44^{ed} \pm 0.30$	$2.14^{h} \pm 0.13$	$28^{d} \pm 0.47$	$2.58^{f} \pm 0.54$	$1.74^{i} \pm 0.05$
10	$68^{ab} \pm 4.47$	$4.64^{cd} \pm 0.29$	$5.42^{e} \pm 0.08$	$50^{c} \pm 7.07$	$4.32^{ed} \pm 0.24$	$2.56^{g} \pm 0.13$
20	$76^{a} \pm 5.48$	$5.88^{a} \pm 0.04$	$8.06^{a} \pm 0.13$	$60^{bc} \pm 7.07$	$4.34^{ed} \pm 0.21$	$4.08^{f} \pm 0.08$
30	$70^{ab} \pm 7.07$	$5.04^{bc} \pm 0.05$	$6.36^{b} \pm 0.11$	$70^{ab} \pm 7.07$	$5.42^{b} \pm 0.11$	$4.52^{e} \pm 0.19$
40	$56^{c} \pm 5.48$	$4.68^{cd} \pm 0.24$	$5.68^{c} \pm 0.13$	$54^{c} \pm 5.48$	$4.08^{e} \pm 0.19$	$3.88^{f} \pm 0.13$
CV%	10.38	5.60	2.72	10.38	5.60	2.72

*NAA =α-naphthalene acetic acid. Values in the same column and variables with different letters are significantly different from each other according to REGWQ at P<0.05.

Figure 1: Acclimatized plantlets. (A) Genotype N52 (B) Genotype N53.

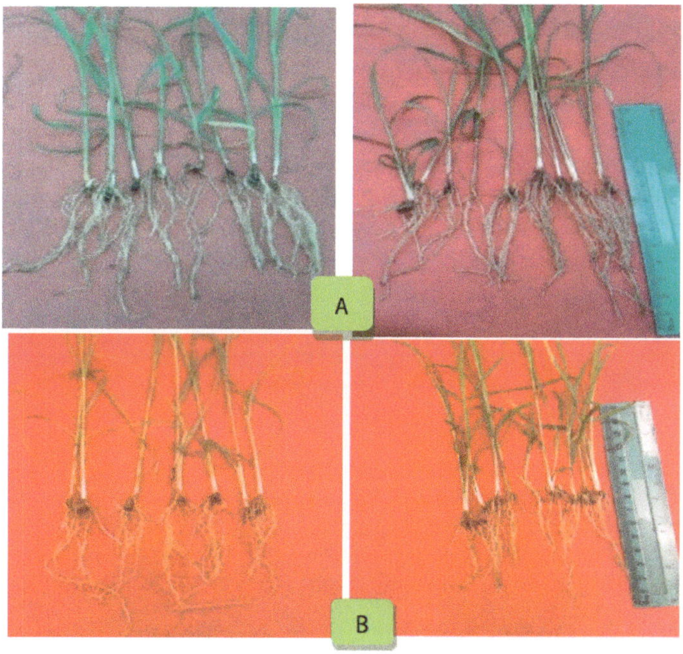

Figure 2: Ex vitro rooting of sugarcane micro-shoots. (A) Genotype N52 at 20 mg/l NAA (B) Genotype N53 at 30 mg/l NAA

guineensis Jacq.). These results indicated the significance of treating of microshoots with plant growth regulators during ex vitro rooting before culturing in soil medium. There was a significant response variation in rooting between the two genotypes. Genotype N52 had the highest (76 %) rooting frequency with a maximum (5.88 cm ± 0.04 cm) average root length and 8.06 ± 0.13 average number of roots per shoot on microshoots dipped in 20 mg/l concentration of NAA (Table 2 and Figure 2A). At the same concentration of NAA, N53 had only 60 % rooting frequency with 4.34 cm ± 0.21 cm average root length and 4.08 ± 0.08 average roots number per shoot. On the other hand, genotype N53 showed a maximum rooting frequency (70 %) with 5.42 cm ± 0.08 cm and 4.52 ± 0.19 numbers of roots per shoot on microshoots treated with 30 mg/l concentration of NAA (Table 2 and Figure 2B). At this concentration of NAA, genotype N52 gave almost equal rooting frequency (70 %) with comparable root length (5.04 cm ± 0.05cm) and higher (6.36 ± 0.11) root number per shoot than N53.The result of this experiment revealed that genotype N52 was more responsive than N53 for different NAA concentrations. The rate of rooting frequency increased from 36 % in control shoots to 76 % when the basal ends of shoots were dipped in a solution of 20 mg/l NAA overnight (Figure 3). Similarly, average root length and average number of roots increased from 4.44 cm ± 0.30cm and 2.14 ± 0.13 to 5.88 ± 0.04cm and 8.06 ± 0.13, respectively, in genotype N52. However, when NAA concentration was further elevated to higher concentration (40 mg/l), rooting frequency, average root length and average number of roots per shoot, reduced significantly to 56 %, 4.68 ± 0.24 cm and 5.68 ± 0.13, respectively. The same trend was observed in genotype N53, in that rooting frequency increased from 28% in control shoots to 70 % in treated shoots with a solution of NAA at 30 mg/l. The average root length and average roots number also increased from 2.58 cm ± 0.54 cm and 1.74 to 5.42 cm ± 0.08cm and 4.52 ± 0.19 respectively, as the concentration of NAA increased from 0.0 mg/l to 30 mg/l but as the concentration of NAA was increased to 40 mg/l the rooting frequency, root length and root number reduced markedly to 54 %, 4.08 cm ±0.19 cm and 3.88 ± 0.13 respectively (Figures 3-5). This reduction in rooting response could be due to the fact that higher concentrations of NAA promote the biosynthesis and buildup of ethylene at the basal end of the shoot, which have inhibitory effect on the overall rooting response of sugarcane microshoots [21].

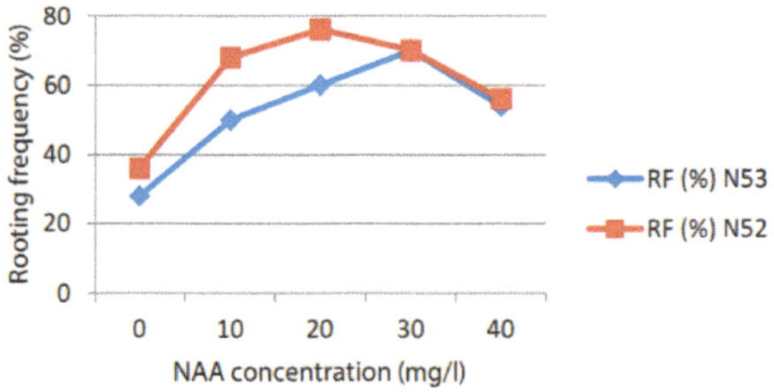

Figure 3: Effect of NAA on rooting frequency of both genotypes.

The result of the present study on genotype N52 were in agreement with earlier results by Pandey et al. [13], who obtained the highest rooting frequency, root length and number of roots per shoot from shoots treated in 20 mg/l of NAA concentration in sugarcane genotype CoS96268. Similarly, Martin [16] obtained an average of 5.6 roots per shoot after the microshoots of Rotula aquatica Lour were dipped in 0.5 mg/l NAA for 25 days. Biradar et al. [20] found best root formation on shoot treated with 2 mM NAA with 80% rooting frequency in oil palm. However, other authors Martin [22] and

Chinnu [23] obtained best result of ex vitro rooting by using IBA. In present study, shoots rooted through this method transplanted to small pots containing soil, sand and farmyard manure (1: l: 1) exhibited 100 % survival in both genotypes.

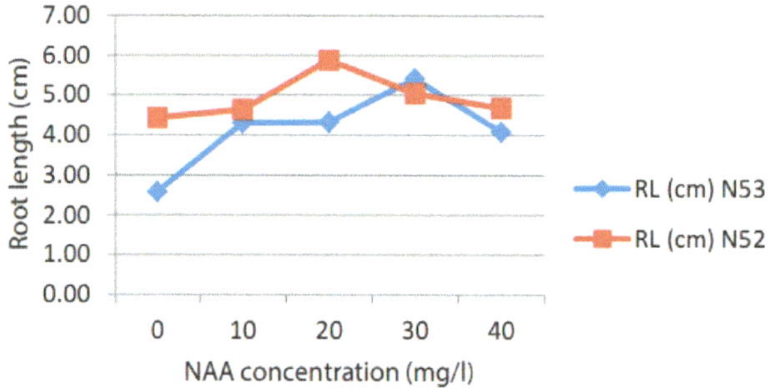

Figure 4: Effect of NAA concentration on root length of both genotypes.

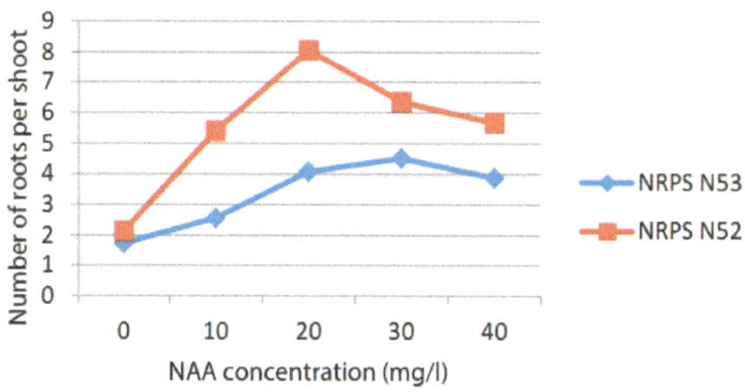

Figure 5: Effect of NAA on number of root per shoot in both genotypes.

CONCLUSION

Generally, it appears that rooting of tissue culture-derived rootless sugarcane plantlets can be induced during ex vitro acclimatization by dipping in the auxin (NAA) solution overnight. From the five concentrations of NAA tested for ex vitro rooting, 20 mg/l NAA was found to be the optimal concentrations for ex vitro rooting of genotype N52. It produced the highest rooting frequency (76 %) with an average of 8.06 roots per shoot while in genotype N53, 30 mg/l NAA gave a maximum of 70 % rooting frequency with 4.52 average root numbers per shoot. Ex vitro rooting reduces the time of acclimatization and labor cost.

ACKNOWLEDGEMENTS

We would like to express our thanks to Ethiopian Sugar Corporation (ESC) for financing the study and Holetta National Agricultural Biotechnology Laboratory (HNABL) for provision of Tissue Culture Laboratory with all required consumable chemicals and facilities.

REFERENCES

1. FAOSTAT (2013) Food and Agriculture Organization of the United Nations Statistics.
2. Ethiopian Sugar Corporation (ESC) (2014) Sweet Newsletter 4:1.
3. Jalaja NC, Neelamathi D, Sreenivasan TV (2008) Micropropagation for Quality Seed Production in Sugarcane in Asia and the Pacific. Food and Agriculture Organization of the United Nations and Asia–Pacific Consortium on Agricultural Biotechnology, Asia Pacific Association of Agricultural Research Institutions.
4. Lorenzo JC, Ojeda E, Espinosa A, Borroto C (2001) Field performance of temporary immersion bioreactor derived sugarcane plant.In Vitro Cellular & Developmental Biology-Plant 37: 803-806.
5. Tilahun M, Mulugeta D, Sharma M (2013) An alternative safer and cost effective surface sterilization method for sugarcane (Saccharum officinarum L.) explants. African J Biotechnol 12: 6282-6286.
6. Belay T, Mulugeta D, Derbew B (2014) Effects of 6-Benzyl aminopurine and Kinetin on in vitro shoot multiplication of sugarcane (Saccharum officinarumL.) varieties.Adv CropSci Tech 2: 129.
7. Dereje S, Kassahun B,Tiliye F (2014) Interaction effect of 6-Benzylaminopurine and Kinetin on in vitro shoots multiplication of two sugarcane (Saccharum officinarum L.) genotypes.Adv Crop Sci Tech 2: 143.
8. Gemechu A, Firew M, Adefris T (2014) Effect of genotype on in vitro propagation of elite sugarcane (Saccharum officinarum L.) varieties of Ethiopian sugar estates. International journal of technology enhancements and emerging engineering research 2.
9. LevaA(2011) Innovative protocol for "ex vitro rooting" on olive micropropagation. Central European Journal of Biology 6: 352-358.
10. HazarikaBN (2006) Morpho-physiological disorder in in vitro culture of plants. SciHort 108: 105-120.
11. Feyissa T,Welander M,Negash L (2007) Genetic stability, ex vitro rooting and gene expression studiesin Hageniaa byssinica.Biologia Plantarum 51: 15-21.
12. Shekafandeh A (2007) Effect of different growth regulators and source of carbohydrates on in and ex vitro rooting of Iranian myrtle. Int J Agricultural Res 2: 152-158.
13. Pandey RN, RastogiJ, Sharma ML, Singh RK (2011) Technologies for cost reduction in sugarcane micropropagation. African J Biotechnol 10: 7814-7819.
14. Bozena B (2001) Morphological and physiological characteristics of micropropagated Strawberry plants rooted in vitro or ex vitro. Sci Hort 89:195–206.
15. Yan H, Liang C, Yang L,Li Y (2010)In vitro and ex vitro rooting of Siratiagrosvenorii, a traditional medicinal plant. Acta physiologiae plantarum 32: 115-120.

16. Martin KP (2003) Rapid in vitro multiplication and ex vitro rooting of Rotulaaquatica Lour, a rare rhoeophytic woody medicinal plant.Plant Cell Report 21: 415–420.

17. Murashige T, and Skoog F (1962) A revised medium for rapid growth and bio assays with tobacco tissue cultures. Physiologia plantarum 15: 473-497.

18. Pathak S, Lal M, Tiwari AK, Sharma ML (2009) Effect of growth regulators on in vitro multiplication and rooting of shoot culture in sugarcane. Sugar Tech 11:86-88.

19. Melaku T, Belayneh A, Kassahun B (2016) In vitro shoot multiplication of elite sugarcane (Saccharum Officinarum L.) genotypes using liquid shake culture system. J Biol Agriculture Healthcare 6:35-40.

20. Sumaryono S, Riyadi I (2011) Ex vitro rooting of oil palm (Elaeisguineensis Jacq.) plantlets derived from tissue culture. Indonesian Journal of Agricultural Science 12: 57-62.

21. Biradar S, Biradar DP, Patil VC, Patil SS, Kambar NS (2009) In vitro plant regeneration using shoot tip culture in commercial cultivar of sugarcane. Karnataka Journal Agricultural Science 22: 21-24.

22. Martin KP (2003) Rapid axillary bud proliferation and ex vitro rooting of Eupatorium triplinerve. Biologia Plantarum 47: 589-591.

23. Chinnu JK, Mokashi AN, Hegde RV, Patil VS, Koti RV (2012) In vitro shoot multiplication and ex vitro rooting of cordyline (Cordyline sp.). Karnataka Journal of Agricultural Science 25: 221-223.

3

CHAPTER

EFFECT OF FODDER RADISH (RAPHANUS SATIVUS L.) GREEN MANURE ON POTATO WILT, GROWTH AND YIELD PARAMETERS

Hayfa Jabnoun-Khiareddine[1], Rania Aydi Ben Abdallah[1,2], Ayed Fakher[3], Mouna Gueddes-Chahed[1], Ahmed Hajlaoui[1], Samir Ben Salem[1], Wissem Ben Dhia[1] and Mejda Daami-Remadi[1]

[1]UR13AGR09- Integrated Horticultural Production in the Tunisian Centre- East, Regional Center of Research on Horticulture and Organic Agriculture, University of Sousse, 4042, Chott-Mariem, Tunisia

[2]National Agronomic Institute of Tunisia, 1082 Tunis Mahrajène, University of Carthage, Tunisia

[3]Technical Center of Organic Agriculture, 4042, Chott-Mariem, Tunisia

ABSTRACT

Potato is threatened by several soil-borne fungi causing wilt and root rots. In this study, two fodder radish (Raphanus sativus L.) (FR) cultivars (cvs. Boss and Defender), used as green manure preceding a potato crop, were evaluated for their suppressive effects against wilt incidence and severity, potato growth and yield as compared to animal manure. The essay was carried out in a completely randomised design with three types of organic amendment and two potato cultivars (cvs. Spunta and Royal). Incidence of potato wilting noted 100 days post planting (DPP) was high, exceeding 70%, for all soil amendments tested. The extent of vascular discoloration varied depending on amendments used where cv. Defender behaved as the control while the highest extent was noted on potato plants grown in cv. Boss amended plots. As compared to animal manure, the application of cvs. Boss and Defender had increased by 48.43 and 41.28% the incidence of vascular discoloration on cv. Spunta, respectively, while on cv. Royal, only cv. Defender had

Citation: Jabnoun-Khiareddine H, Abdallah RAB, Fakher A, Gueddes-Chahed M, Hajlaoui A, et al. (2016) Effect of Fodder Radish (Raphanus sativus L.) Green Manure on Potato Wilt, Growth and Yield Parameters. Adv Crop Sci Tech 4:211. doi:10.4172/2329-8863.1000211.

reduced this parameter by 16.32%. Fungal isolations performed from roots and stems revealed the involvement of several soil-borne pathogens in the recorded plant wilting. Soil manuring using cvs. Boss and Defender FR resulted in significant increment in average stem number per plant and aerial part fresh weight by 22.79 and 21.32% and by 34.62 and 27.03%, respectively, as compared to animal manure. At 100 DPP, potato root fresh weight increase by 8.7 and 33.49% was noted on cv. Spunta compared to 30.34 and 23.48% recorded on cv. Royal. Potato tuber yield was improved by 38.28 and 10.7% and by 28.44 and 27.62% in cvs. Spunta and Royal, respectively, relative to animal manure. The use of FR as green manure may be implemented in the integrated management of soil-borne diseases for the enhancement of potato yield.

INTRODUCTION

In Tunisia, potato (Solanum tuberosum L.) is one of the strategic crops which occupies about 16% of all Tunisian cultivated areas [1-3]. Potato production is concentrated in relatively small diversified farms where potato is grown for several years in the same fields leading to the build-up of many soil-borne pathogens [4]. In fact, soil-borne diseases are persistent, recurrent problems in potato production, resulting in decreased plant growth and vigor, lower tuber quality incited mainly by skin blemishing pathogens and reduced yield [5,6]. Of particular concern throughout major potato producing regions in Tunisia, due to their constant presence, are Fusarium wilt caused mainly by Fusarium oxysporum f. sp. tuberosi [7,8], Verticillium wilt incited by Verticillium dahliae [9], Black dot caused by Colletotrichum coccodes [10], and Rhizoctonia canker and black scurf incited by Rhizoctonia solani [2,3,11]. In Tunisia, Fusarium wilt has become in last decade as one of the most serious potato diseases responsible for plant stunting, severe leaf yellowing, vascular discoloration and subsequent plant wilting and death [7,8]. Verticillium wilt is a common potato disease causing premature senescence leading to 30-50% yield losses [9,12]. For C. coccodes, symptoms include sloughing of the root cortex, brown lesions on roots and blemishes on tubers [13]. This pathogen can cause premature foliage death and yield losses were reported to be as much as 30% [14,15]. R. solani affects potato development from emergence to harvest and typical symptoms include death of preemerging sprouts, cankers on underground stem parts and stolons, a diminished root system, and the formation of sclerotia on progeny tubers which leads to lowered tuber yield and quality [16]. These pathogens are often observed as mixed infections and are also associated with other soil-borne bio-aggressors such as nematodes, which lead to exacerbated potato early dying syndrome. This increased prevalence and severity of soil-borne fungal diseases was due to the absence of resistant potato cultivars since the most grown cultivars exhibited varying degree of susceptibility to all these pathogens [17-21].

Several strategies have been used, in Tunisia and all over the world, to control these potato diseases, such as soil solarisation, long-term rotations, biological control, host resistance, etc. However, serious losses still occur due mainly to pathogen's long-lived resting structures released in the soil such as Verticillium microsclerotia, Fusarium chlamydospores and C. coccodes and R. solani sclerotia formed in senescing and dead potato roots, stems, stolons and tubers. Hence, a crucial factor in the management of diseases caused by these pathogens is to reduce their inoculum level below the critical threshold level before a susceptible crop is planted [22].

In the last few years, old practices such as incorporation of green manure, animal manure, compost, seed meals and other types of organic amendments have been used to control soil-borne pathogens such as Phytophthora spp., Fusarium spp., Verticillium spp., Rhizoctonia spp, and Sclerotinia spp. [23-27] leading to interesting and promising results [28]. For example, the incidence of Verticillium wilt of potato has been reduced using pea (Pisum sativum), oat (Avena sativa), broccoli (Brassica oleracea), sudan-grass (Sorghum vulgare) and corn (Zea mays) as green manures [29-31], and addition of animal manures [32,33].

However, some studies indicate that the effectiveness of organic amendments including Brassica residues is variable and, in some cases, can even enhance disease severity [34]. This negative effect of organic amendments may be attributed to increased pathogen inoculums' potential, as the decaying material may serve to sustain saprophytic growth of plant pathogens [35], or carried out by the Brassica amendments themselves [34]. Furthermore, increases in tuber yield using green manures crops have been inconsistent, despite the beneficial disease suppressive effects [36-38].

Green manures based on Brassicaceae species, incorporated into soil when still green or soon after maturity, are shown able to improve the soil's physical, chemical, or biological properties and thereby to increase the succeeding crop's yield, quality, or both [39]. Among these Brassicas, fodder radish (Raphanus sativus L.) (FR) presents many unique characteristics such as its relatively high tissue phosphor concentration, rapid dry matter accumulation in the fall, and rapid residue decomposition in the spring [40].

FR and other Brassicas produce isothiocyanates as a by-product of glucosinolate break-down, and often have a metham sodium like biofumigant action to suppress weeds, nematodes and other soil-borne pests and diseases [28,37,38,41-44]. Furthermore, addition of organic matter increases resistance of the individual plant as a result of uptake of phenols, phenolic and other compounds, such as salicylic acid, which have an antibiotic effect and also work directly on pathogens [45]. Green manures may also affect soil-borne pathogens indirectly by influencing indigenous microbial populations. In fact, incorporation of soil amendments leads to increased soil microbial activity and diversity [46] as well as density of bacteria [47], fluorescent Pseudomonas spp. [32,47,48], streptomycetes, and other actinomycetes [47,49] in soil.

The use of Brassica spp. and related plants as green manure crops has been receiving increased attention in recent years for their ability to reduce multiple soil-borne potato diseases [50]. Therefore, the current research was established to evaluate the effect of two FR cultivars used as green manure crop, preceding potato growing, for their ability to reduce potato soil-borne diseases in naturally infested soil in Chott- Mariem region, Centre East Tunisia, and their efficacy in increasing potato growth and tuber yield.

MATERIALS AND METHODS

Plant Material

Potato cvs. Spunta and Royal seed tubers were kindly provided by the Technical Center of Potato and Artichoke, Essaïda, Tunisia. Cv. Spunta is most grown, highly susceptible to vascular wilts and lowyielding, whereas cv. Royal is recently introduced in Chott-Mariem region. Before use, tubers were superficially disinfected with a 10% sodium hypochlorite solution during 5 min, rinsed with tap water and air dried. They were kept two weeks under 15-20°C, 60-80% relative humidity and natural room light for pre-germination.

Two FR (oilseed radish) cultivars: cvs. Boss and Defender were kindly provided by the General Directory of the Protection and the Control of the Agricultural Product Quality, Ministry of Agriculture, Tunisia, for the management of soil-borne bio-aggressors. These FR cultivars were used as green manure crops preceding potato cropping.

Field site, Crop Sequences and Design

Field trial was conducted in 2011-2012 at the experimental farm of the Regional Center of Research on Horticulture and Organic Agriculture in Chott-Mariem region. This site is under conventional farming system and has a history of potato and other vegetables production practices. The soil has a sandy clay texture (Organic matter 76 g/kg at 0-20 cm depth) and has also a long history of potato soilborne fungal diseases such as Verticillium and Fusarium wilts, Black dot and Rhizoctonia stem canker.

Prior to planting potato, radishes were sown at a rate of 25 kg/ ha and grown for approximately nine weeks in the fall (September 12- November 19, 2011). These green manure crops were cut down at flowering stage and mechanically incorporated into soil to a depth of 15-20 cm, between 20 to 23 November by rotovating. The fresh matter (T/ha) of green manures incorporated into the soil was 6 for cv. Boss and 9 for cv. Defender.

The non-radish control treatment consisted of the standard organic amendment, for that site, which was animal manure which is composed of bovine wastes and was applied at a rate of 50 T/ha two weeks before planting potato.

At multi-germ stage, potato seed tubers were planted on December 12, 2012 in radish and animal manure-amended plots consisting of two 40 m long rows. Seed rows were 1 m apart, with two rows per plot, and within row spacing of 0.4 m.

The trial was set up in a completely randomized design with three replications. For each organic amendment and in each replicate or mini plot, thirty potato plants were used per potato cultivar.

Disease Incidence and Severity

Observational wilt incidence was determined late in the growing season, 100 days post planting (DPP), as the proportion of plants, per plot and per potato cultivar, exhibiting typical early dying symptoms such as leaf chlorosis, wilting or death. For the assessment of disease severity, ten randomly selected plants, with wilt symptoms, were dug from the rows of each organic amendment mini plot. For each plant, stems were longitudinally cut and visually examined for the presence of vascular discoloration. The extent of this vascular discoloration was noted per plant and its incidence represents the percentage of stems per plant showing discolored vessels.

For each organic treatment, plant stems and roots were mixed, surface sterilized in 0.5% NaOCl, rinsed in sterile water, cut into 3-5 mm-fragments, and plated on Potato Dextrose Agar (PDA) medium for the isolation of fungal pathogens involved in wilting symptoms.

The isolation frequency for each pathogen was calculated as the percentage of root or stem fragments showing pathogen growing colonies relative to the total number of stem fragments plated on PDA.

Potato Growth and Yield Parameters

During this essay, cultural practices involving potato fertility, irrigation and pesticide applications were the most commonly used in the region, as recommended by the Technical Center of Potato and Artichoke for potato production guidelines. At 100 DPP, the same ten randomly selected plants, as described above for disease assessment, served for the measurement of roots, aerial parts and tubers' fresh weights together with average stem number per plant.

Statistical Analyses

For all parameters measured (disease incidence and severity, stem number per plant, root, aerial part and tuber fresh weight per plant, as well as the isolation frequency per pathogen), statistical analyses were performed following a completely randomised design with three replications where treatments (FR cv. Boss, FR cv. Defender and the animal manure) and potato cultivars (Spunta and Royal) represented the two fixed factors. Thirty potato plants were used per individual treatment. Mean separation was carried out with Fisher's protected LSD or Duncan's Multiple Range test (at $P \leq 0.05$). Statistical analyses were performed using SPSS software version 16.

RESULTS

Effect of the Organic Amendments Tested on Potato wilt Incidence

Visible potato wilt incidence estimated based on the presence of wilt and foliage chlorosis noted at the end of the growing season (100 DPP) was high, exceeding 70%, for all soil amendments tested. In fact, this incidence was not significantly (at $P \leq 0.05$) affected neither by soil amendments and potato cultivars nor by their interaction (Table 1). However, potato plants grown following FR cv. Boss green manure showed higher wilt incidence (85%), even statistically insignificant, than following cv. Defender (77%) or using animal manure (73%) as control amendment and this whatever the potato cultivar used. Furthermore, for all soil amendments tested, wilt incidence noted on cv. Spunta plants tended to be higher than that recorded on cv. Royal, even if statistically insignificant.

Table 1: Effect of Boss and Defender fodder radish cultivars used as green manure preceding potato growing, compared to animal manure, on wilt incidence (%) on two potato cultivars noted 100 days post-planting

Organic amendments/ Potato cultivars	**cv. Spunta**	**cv. Royal**	**Average incidence per organic amendment[a*]**
Folder radish cv. Boss	94.78	75.67	85.22 a
Folder radish cv. Defender	81.78	71.33	76.55 a
Animal manure	70.56	75.67	73.11 a
Average incidence per potato cultivar[b*]	82.37 a	74.22 a	

a Mean wilt incidence per organic amendment for potato cultivars combined.
b Mean wilt incidence per potato cultivar for all organic amendments combined.
*For organic amendments and potato cultivars tested, values followed by the same letter are not significantly different according to Duncan's Multiple Range test at $P \leq 0.05$.
Wilt incidence was determined as the proportion of plants with leaf chlorosis, wilting or death in each plot for each potato cultivar. Thirty plants were used per individual treatment.

Table 2: Effect of Boss and Defender fodder radish cultivars used as green manure preceding potato growing, compared to animal manure, on the extent of vascular discoloration on two potato cultivars noted 100 days post-planting

Organic amendments/ Potato cultivars	**cv. Spunta**	**cv. Royal**	**Average extent per organic amendment[a*]**
Folder radish cv. Boss	6.58	5.69	6.15 a
Folder radish cv. Defender	4.59	2.65	3.62 b
Animal manure	1.94	3.10	2.52 b
Average extent per potato cultivar[b*]	4.37 a	3.81 a	

a Mean vascular discoloration extent per organic amendment for potato cultivars combined.
b Mean vascular discoloration extent per potato cultivar for all organic amendments combined.
*For organic amendments and potato cultivars tested, values followed by the same letter are not significantly different according to Duncan's Multiple Range test at $P \leq 0.05$.Ten plants were used per individual treatment.

Table 3: Effect of Boss and Defender fodder radish cultivars used as green manure preceding potato growing, compared to animal manure, on the incidence of stem vascular discoloration (%) on two potato cultivars noted 100 days post-planting

Organic amendments/ Potato cultivars	cv. Spunta	cv. Royal	Average incidence per organic amendment[a*]
Folder radish cv. Boss	58.5	69.23	63.87 a
Folder radish cv. Defender	51.38	48.11	49.75 b
Animal manure	30.16	57.5	43.83 b
Average incidence per potato cultivar[b*]	46.68 b	58.28 a	

LSD (Organic amendments x potato cultivars)=24.21 at P ≤ 0.05.
a Mean stem vascular discoloration incidence per organic amendment for potato cultivars combined.
b Mean stem vascular discoloration incidence per potato cultivar for all organic amendments combined.
*For organic amendments and potato cultivars tested, values followed by the same letter are not significantly different according to Duncan's Multiple Range test at P ≤ 0.05.
The incidence of stem vascular discoloration represents the percentage of stems per plant showing discolored vessels. Ten plants were used per individual treatment.

Effect of the Organic Amendments Tested on Potato wilt Severity

Data presented in Table 2 showed that wilt severity, estimated via the extent of vascular discoloration, noted on potato stems varied significantly (at $P \leq 0.05$) depending only on the organic amendments tested. The highest extent, of 6.15 cm from the collar, was recorded on potato plants grown in cv. Boss-amended plots, which is 58.91% higher than that noted on plants grown following animal manure. However, plants from cv. Defender-treated plots showed wilt severity statistically comparable to that of plants grown in animal-manured plots.

Potato cv. Spunta plants showed, for all soil amendments tested combined, 12.73% higher vascular discoloration extent, even statistically insignificant, than that noted on cv. Royal plants.

ANOVA analysis revealed that vascular discoloration incidence, noted on potato plants 100 DPP, varied significantly (at $P \leq 0.05$) depending on soil treatments and potato cultivars; a significant interaction was noted between both fixed factors. In fact, data given in Table 3 indicated that this incidence ranged from 30.16%, noted on cv. Spunta plants grown in animal manure-amended plots, to 69.23%, recorded on cv. Royal grown following cv. Boss-treated plots. As compared to animal manure, green amendments using cvs. Boss and Defender FR had increased the incidence of vascular discoloration per plant by 48.43 and 41.28% on cv. Spunta, respectively. However, on cv. Royal, soil amendment using cv. Defender had reduced this parameter by 16.32%, as compared to animal manure, contrarily to cv. Boss, which increased this incidence by 16.94%. Overall and whatever the organic amendment tested (combined data of the three organic amendments tested), the incidence of vascular discoloration noted on cv. Spunta plants was 19.9% lesser than that recorded on cv. Royal plants.

Table 4: Effect of Boss and Defender fodder radish cultivars, used as green manure preceding potato growing, on the isolation frequency of soil-borne pathogens from roots, as compared to animal manure, noted 100 days post-planting

Organic amendments/ Pathogens	cv. Boss	cv. Defender	Animal manure	Average isolation frequency (%) per fungus[a*]
Fusarium oxysporum	35.00	28.33	53.33	38.89 a
F. solani	12.50	11.67	12.83	12.33 b
Verticillium dahliae	10.83	11.67	11.67	11.39 b
Rhizoctonia solani	7.50	15.83	24.17	15.83 b
Colletotrichum coccodes	32.50	43.33	25.83	33.89 a

a Mean isolation frequency per pathogen for all organic amendments and potato cultivars combined.

*For fungal pathogens, values followed by the same letter are not significantly different according to Duncan's Multiple Range test at P ≤ 0.05.

Table 5: Effect of Boss and Defender fodder radish cultivars, used as green manure preceding potato growing, on the isolation frequency of soil-borne pathogens from stems, as compared to animal manure, noted 100 days post-planting

Organic amendments/ Pathogens	cv. Boss	cv. Defender	Animal manure	Average isolation frequency (%) per fungus[a*]
Fusarium oxysporum	57.50	55.83	47.50	53.61 a
F. solani	10	0.83	4.17	5 b
Verticillium dahliae	10.83	19.17	15	15 b
Rhizoctonia solani	5.83	9.17	15	10 b
Colletotrichum. coccodes	8.33	17.50	12.50	12.78 b

a Mean isolation frequency per pathogen for all organic amendments and potato cultivars combined.

*For fungal pathogens, values followed by the same letter are not significantly different according to Duncan's Multiple Range test at P ≤ 0.05.

Effect of the Organic Amendments Tested on the Isolation Frequency of soil-Borne Pathogens

At the end of the growing season (100 DPP), fungal isolations performed from potato plants grown in organically amended plots and exhibiting wilt symptoms revealed the involvement of several soilborne fungal pathogens in the observed signs. In fact, the isolation frequency of these pathogens was not significantly influenced nor by potato cultivars and organic amendments, nor by their interaction (at $P \leq 0.05$). As shown in Tables 4 and 5, F. oxysporum and C. coccodes were most frequently isolated from roots of wilting plants, followed by R. solani, F. solani, and V. dahliae whereas the predominant fungi recovered from stems showing vascular discoloration and cankers were F. oxysporum, V. dahliae, C. coccodes, R. solani, and to a lesser extent F. solani.

Effect of the Organic Amendments Tested on Potato Growth and Production Parameters

Average stem number: Results illustrated in Table 6 indicated that the average stem number noted on potato plants at the end of the growing season (100 DPP) varied depending on organic amendments only. In fact, soil treatment using FR cvs. Boss and Defender led to significant increase in average stem number per plant by 22.79 and 21.32%, respectively, as compared to animal manure

Root fresh weight: Potato root fresh weight was significantly affected by the organic amendments tested but was not influenced with potato cultivars used; however, a significant interaction (at $P \leq 0.05$) was noted between both fixed factors (Table 7). In fact, as indicated in this table, FR cvs. Defender and Boss had increased this parameter by 8.7 and 33.49% on cv. Spunta compared to 30.34 and 23.48% recorded on cv. Royal, respectively, as compared to animal manure.

Table 6: Effect of Boss and Defender fodder radish cultivars, used as green manure preceding potato growing, as compared to animal manure, on mean stem number per plant on two potato cultivars tested noted 100 days post-planting

Organic amendments/ Potato cultivars	cv. Spunta	cv. Royal	Average stem number per organic amendment[a*]
Fodder radish cv. Boss	3.77	3.40	3.58 a
Fodder radish cv. Defender	3.57	3.47	3.52 a
Animal manure	3.10	2.43	2.77 b
Average stem number per potato cultivar[b*]	3.48 a	3.10 a	

a Mean stem number per organic amendment for potato cultivars combined.
b Mean stem number per potato cultivar for all organic amendments combined.
*For organic amendments and potato cultivars tested, values followed by the same letter are not significantly different according to Duncan's Multiple Range test at $P \leq 0.05$. Ten plants were used per individual treatment.

Aerial part fresh weight: Data given in Table 8 revealed that only the organic amendments tested had significantly impacted the aerial part fresh weight noted on potato plants at the end of the growing season (100 DPP). In fact, green manuring using FR cvs. Boss and Defender had significantly enhanced potato aerial part fresh weight by 34.62 and 27.03%, respectively, relative to animal manure. This parameter was 12.57% higher on potato cv. Royal plants, even statistically insignificant, than that noted on cv. Spunta plants.

Table 7: Effect of Boss and Defender fodder radish cultivars, used as green manure preceding potato growing, as compared to animal manure, on root fresh weight (g) of two potato cultivars noted 100 days post-planting

Organic amendments/ Potato cultivars	cv. Spunta	cv. Royal	Average fresh weight per organic amendment[a*]
Fodder radish cv. Boss	27.87	23.70	25.78 a
Fodder radish cv. Defender	20.30	26.03	23.17 a
Animal manure	18.53	18.13	18.33 b
Average fresh weight per potato cultivar[b*]	22.23 a	22.62 a	

LSD (Organic amendments × potato cultivars)=6.94 g at P ≤ 0.05.
a Mean root fresh weight per organic amendment for potato cultivars combined.
b Mean root fresh weight per potato cultivar for all organic amendments combined.
*For organic amendments and potato cultivars tested, values followed by the same letter are not significantly different according to Duncan's Multiple Range test at P ≤ 0.05.Ten plants were used per individual treatment.

Table 8: Effect of Boss and Defender fodder radish cultivars, used as green manure preceding potato growing, as compared to animal manure, on aerial part fresh weight (g) of two potato cultivars noted 100 days post-planting

Organic amendments/ Potato cultivars	cv. Spunta	cv. Royal	Average fresh weight per organic amendment[a*]
Fodder radish cv. Boss	126.90	148.50	137.70 a
Fodder radish cv. Defender	108.60	138.13	123.37 a
Animal manure	92.03	88.00	90.02 b
Average fresh weight per potato cultivar[b*]	109.18 a	124.88 a	

a Mean aerial part fresh weight per organic amendment for potato cultivars combined.
b Mean aerial part fresh weight per potato cultivars for all organic amendments combined.
*For organic amendments and potato cultivars tested, values followed by the same letter are not significantly different according to Duncan's Multiple Range test at P ≤ 0.05.Ten plants were used per individual treatment.

Table 9: Effect of Boss and Defender fodder radish cultivars, used as green manure preceding potato growing, as compared to animal manure, on tuber fresh weight (g) of two potato cultivars noted 100 days post-planting

Organic amendments/ Potato cultivars	cv. Spunta	cv. Royal	Average fresh weight per organic amendment[a*]
Fodder radish cv. Boss	989.90	727.53	858.72 a
Fodder radish cv. Defender	684.10	719.33	701.72 b
Animal manure	610.93	520.60	565.77 c
Average fresh weight per potato cultivar[b*]	761.64 a	655.82 b	

LSD (Organic amendments x potato cultivars)=184.58 g at P ≤ 0.05.
a Mean tuber fresh weight per organic amendment for potato cultivars combined.
b Mean tuber fresh weight per potato cultivar for all organic amendments combined.
*For organic amendments and potato cultivars tested, values followed by the same letter are not significantly different according to Duncan's Multiple Range test at P ≤ 0.05. Ten plants were used per individual treatment.

Tuber fresh weight: As shown in Table 9, tuber fresh weight per plant varied significantly (at P ≤ 0.05) depending on amendments used and on potato cultivars tested; a significant interaction was also noted between both fixed factors. In fact, tuber yield per plant, recorded at the end of the growing season (100 DPP), ranged between 990 g, recorded on cv. Spunta plants grown in cv. Boss-amended plots, and 520.6 g, noted on cv. Royal plants grown following animal manure. In fact, soil treatment using FR cvs. Boss and Defender resulted in 38.28% and 10.7% higher tuber fresh weight on cv. Spunta compared to 28.44% and 27.62% increase recorded on cv. Royal, respectively, as compared to animal manure. Furthermore and whatever the potato cultivar used, manuring with FR cvs. Boss and Defender resulted in 34.11 and 19.37% higher tuber yield per plant than animal manure. Furthermore, cv. Spunta yielded significantly 13.89% higher tuber fresh weight than cv. Royal for all amendments combined.

DISCUSSION

In the present study, two cultivars of FR, grown as green manures preceding potato growing, were assessed, in comparison with animal manure, for their ability to reduce the incidence of potato soil-borne fungal pathogens and to ultimately increase crop growth and yield.

Green manuring with FR cvs. Boss and Defender, before planting potato as main crop, resulted at the end of the growing season in high wilt incidence levels in the treated plots as well as in animal manured ones. However, wilt severity, estimated via vascular discoloration extent, noted following cv. Defender-amendment was comparable to that recorded on plants grown in animal manured plots. In contrast, 58.91% higher vascular discoloration extent, relative to animal manure, was noted on plants from cv. Boss-amended plots.

These results indicated that single application of FR cvs. Boss and Defender did neither reduce wilt incidence nor wilt severity better than animal manure under our field conditions. In fact, the site chosen in this investigation is known to be highly infested by soil-borne and vascular pathogens causing yearly severe losses in terms of quantity and quality of potato tubers (unpublished data). Therefore, the observed wilt incidence noted on potato plants is the coupled effect of the root infecting pathogens,

C. coccodes, R. solani and F. solani, together with the vascular ones namely F. oxysporum f. sp. tuberosi and V. dahliae. In fact, our results are consistent with many other findings showing that manuring soil with Brassicaceae species have not shown always efficacy in reducing disease severity and soil-borne inoculums in the soil. In this sense, Neubauer et al. [51] find that soil amendments using R. sativus cultivars such as cv. Defender are less effective than B. juncea in reducing viable microsclerotia. Furthermore, the incorporation of broccoli (B. oleracea) does not significantly reduce inocula of F. oxysporum f. sp. asparagi, R. solani, V. dahliae, and Fusarium spp. [22,52]. In this regard, Hartz et al. [53] mention that soil populations of V. dahliae and Fusarium spp. are unaffected by Brassica spp.-based green manures and there are no evidence of soil-borne disease suppression on subsequent tomato crops. Results from the current study are also in line with those of Davis et al. [54] revealing that two consecutive years of sudangrass, oat, or rye green manures does not reduce inoculum of V. dahliae. In the same sense, Geary et al. [55] find that oilseed radish, mustard, and canola exhibit limited effects on the severity of onion pink root caused by Phoma terrestris, and are not a viable option for this disease control. Moreover, Molina et al. [56] mention that soil treatments with green manures based on fall rye (Secale cereale L.) and sorghum-sudangrass hybrid (Sorghum bicolor L. Moench, 'Super Su 22') increase inoculum density as well as potato early dying incidence.

This lack of effectiveness of green manure-based treatments in decreasing wilt incidence and severity may be attributed to several factors such as the limited number of green manure cycles or insufficient physical disruption of plant tissues during plow down that lead to low release of toxic compounds, mainly glucosinolates, as previously suggested [12,57]. These factors are relatively important for long-term survival structures of plant pathogens in soil, particularly, microsclerotia of V. dahliae [12]. In fact, glucosinolate concentrations and the resulting production of different forms of isothiocyanates vary greatly among Brassica species and even among cultivars within each species [58], and are also affected by environmental conditions and plant development [59,60]. This could explain in parts the lack of efficacy in reducing potato diseases in the present work, as FR is known to have moderate glucosinolate content [61] as compared to other Brassicaceae species. In addition, in the field, the expected isothiocyanates concentrations are once more lower, because of a poor release efficiency of these compounds due to incomplete tissue maceration with a mulching implement.

Other studies indicate that the negative effect of these amendments may result from increased pathogens' inoculum potential when the substrate serves to sustain saprophytic growth of plant pathogens [35] or from the increase of inoculum potential of pathogens carried out by the Brassica-based amendments themselves [34].

Nevertheless, the effectiveness of organic amendments including Brassicaceous residues has been shown to be variable [62]. In fact, the adoption of a green manure rotation crop has been associated with significant decreases in severity and incidence of Verticillium wilt, black scurf and stem canker [37,38,63], as well as common scab [37,38]. Larkin and Griffin [37] report that Brassicaceae crops, including radish, canola, rapeseed, turnip, yellow mustard, and Indian mustard are able to reduce inoculum levels of R. solani by 20-56% in greenhouse tests and to decrease subsequent potato seedling disease by 40-83%. Furthermore, oilseed radish incorporated earlier as a green manure has been observed to reduce Rhizoctonia Root and Crown Rot in sugarbeet and to decrease population of V. dahliae to a greater degree [64,65]. In addition, fodder and oilseed radishes have also shown efficacy in suppressing many soil-borne bio-agressors such as nematodes and bacteria. It was found that two new oil radish varieties Defender and Comet have significantly lowered population of Heterodera schachtii by 95%. Hafez and Sundararaj [66] consider that cv. Defender is the most economically and highly suitable cultivar for the management of sugar beet cyst nematode. Moreover, significant reduction and long-term elimination of bacterial wilt caused by Ralstonia solanacearum from the soil was achieved through incorporating especially radish or mustard plants in large amounts into the soil immediately before planting tomatoes which led to 50-70% lower bacterial wilt incidence [67].

In this regard, Davis et al. [68] mention that a single green manure of sweet corn followed by two consecutive potato cropping years, has suppressed by 60-70% Verticillium wilt and increased yields and this effect has occurred even though soil-borne V. dahliae inoculum levels have augmented by more than 4-fold from 45 to 182 cfu/g of soil. They also found that, although these treatments show no direct effect on V. dahliae soil populations, the colonization of V. dahliae on potato feederroots and in potato tissue of stem apices are reduced.

These inconsistencies in reducing microbial populations using Brassicaceae-based amendments have been attributed to the use of different species, physical environments and target organisms [52], soil temperature [69], and amount of crop residue incorporated [22]. In general, the mechanisms involved in disease control are multiple and can vary with each pathosystem [24].

In the current study, manuring soil with FR cvs. Boss and Defender, at a rate of 25 t/ha, had significantly enhanced all potato growth and yield parameters as compared to animal manure. In fact, potato average stem number per plant and aerial part fresh weight were respectively increased by 22.79 and 21.32% and by 34.62 and 27.03%. Furthermore, roots fresh weight were augmented by 8.7 and 33.49% on cv. Spunta compared to 23.48 and 30.34% recorded on cv. Royal, following FR cvs. Boss and Defender. In this sense, several studies indicate the positive effect on plant's biomass produced in forage radish-amended soils such as for Lactuca sativa, Lupinus nanus, and Beta vulgaris [64,70,71]. In this regard, Subbarao and Hubbard [72] find that the incorporation of broccoli residue results in consistently taller cauliflower plants with greater root and shoot biomass.

Moreover, in the present study, potato growth improvement achieved following single application of cvs. Boss and Defender as green manures, resulted in enhancement of tuber yield per plant by 38.28 and 10.7% noted in cv. Spunta and by 28.44 and 27.62% recorded on cv. Royal, respectively, as compared to animal manure. This increase in potato tuber yield using Brassicaceous species green manures was reported in several previous studies [30,38,54]. In this regard, Lehrsch and Gallian [64] find that planting radish as a green manure in the fall, prior to planting potato the following spring, leads to yield improvement in the grown potato as well as in the subsequent sugar beet crops. It has been proved that amongst six varieties of green manure crops, maximum beet yield was obtained from FR cv. Defender planted plots [67]. Our results are also in line with those of Lazarovits [73] showing that incorporating millet over 3 years at two potato field sites, lead to 20% increased potato yields at one site and 50% at the other, but little decrease in Verticillium wilt incidence was observed.

Interesting results from the present study showed that even though disease reductions are minimal, yield increases due to incorporation of radish biomass are significant and this could be explained by the improved nutrient availability as mentioned in several studies [45,74,75]. In this sense, Lehrsch and Gallian [64] relate the increase in the yield and quality of sugar beet, after fall-incorporated radish biomass, to soil physical and hydraulic properties improvement. In this regard, Schomberg et al. [76] report that oilseed radish grows rapidly in the fall and spring and can scavenge significant quantities of N. Furthermore, as oilseed radish produced a significant amount of biomass in the fall and early spring, it has been suggested that it could be useful in rotations where earlier planting dates are desired and for preventing leaching of residual N [76]. In the same sense, Talgre et al. [77] find that among all the Brassicaceae tested, the most effective ones were FR and white mustard which produce the highest biomass and therefore drove more nutrients into the soil. Many studies have also related the disease suppressive effect of green manure to improved soil fertility (mainly NPK contents) [12,78] and soil quality [31]. In fact, green manures and organic amendments are commonly used in crop production systems to increase soil nutrient availability, organic nitrogen and organic matter, which are associated with higher tuber yields [19]. Furthermore, the efficacy of organic amendments is soilspecific and very much rate-dependent. In this sense, Ochiai et al. [31] find that the quantity and not the chemistry of organic inputs is the critical factor for disease suppression.

In the present study, the increase in potato growth and yield seem to be more pronounced using cv. Boss than cv. Defender even though this last cultivar have been incorporated at the rate of 9 kg/ha compared to 6 kg/ha for the former. A possible explanation may involve the beneficial microbial population composition which may be more stimulated by residues' allelochemicals from cv. Boss than those from cv. Defender. Furthermore, cv. Spunta yielded 13.89% significantly higher tuber fresh weight than cv. Royal, for all amendments combined which demonstrated the usefulness of these soil treatments for tuber yield increase in the most grown cultivar in the region even under high soil-borne inoculums levels.

CONCLUSION

Single application of FR cvs. Boss and Defender resulted in a significant growth and yield improvement of two potato cultivars, Spunta and Royal, comparatively to animal manure which is the most used organic amendment in the region, but did not result in significant decrease in wilt diseases' incidence noted at the end of the growing season. Therefore, additional research is needed to verify if multiple seasons together with higher rates of green manures could be associated with significant decreases in wilt incidence and severity on potato and to identify which mechanisms are related with yield improvement. Further studies are also required to determine the effect of these green manures in reducing the pathogens' inoculum densities and enhancing the activity of beneficial micro-organisms in the soil thus improving its fertility.

Furthermore, green manures, like other types of cultural practices, are best implemented as an important component of an integrated disease management program and not as the only control means for soil-borne diseases. Thus, combining green manures with other cultural, biological or chemical approaches can substantially increase disease control and help in achieving greater sustainability and improving soil fertility and subsequently crop production.

REFERENCES

1. Anonymous (2015) Groupement Interprofessionnel des légumes, filières des légumes.
2. Djébali N, Tarhouni B (2010) Field study of the relative susceptibility of eleven potato (Solanum tuberosum L.) varie ties and the efficacy of two fungicides against Rhizoctonia solani attack. Crop Prot 29: 998-1002.
3. Mrabet M, Elkahoui S, Tarhouni B, Djebali N (2015) Potato seed dressing with Pseudomonas aeruginosa strain RZ9 enhances yield and reduces black scurf. Phytopathol Medit 54: 265-274.
4. Triki MA, Priou S, El Mahjoub M (2001) Effects of soil solarization on soil-borne populations of Pythium aphanidermatum and Fusarium solani and on the potato crop in Tunisia. Potato Res 44: 271-279.
5. Larkin RP (2008) Relative effects of biological amendments and crop rotations on soil microbial communities and soilborne diseases of potato. Soil Biol Biochem 40: 1341-1351.
6. Larkin RP, Tavantzis S (2013) Use of biocontrol organisms and compost amendments for improved control of soilborne diseases and increased potato production. Am J Potato Res 90: 261-270.
7. Daami-Remadi M, El Mahjoub M (2004) Emergence en Tunisie de Fusarium oxysporum f. sp. tuberosi agent de flétrissure vasculaire des plantes et de pourriture sèche des tubercules de pomme de terre. EPPO Bull 34: 407-411.
8. Ayed F, Daami-Remadi M, Jabnoun-Khiareddine H, El Mahjoub M (2006) Potato Vascular Fusarium wilt in Tunisia: Incidence and biocontrol by Trichoderma spp. Plant Pathol J 5: 92-98.
9. Daami-Remadi M, Jabnoun-Khiareddine H, Ayed F, El Mahjoub M (2011) Comparative aggressiveness of Verticillium dahliae, V. albo-atrum and V. tricorpus on potato as measured by their effects on wilt severity, plant growth and subsequent yield loss. Funct Plant Sci Biotechnol 5 (Special Issue): 1-8.
10. Daami-Remadi M, Bouallègue R, Jabnoun-Khiareddine H, El Mahjoub M (2010) Comparative aggressiveness of Tunisian Colletotrichum coccodes isolates on potato assessed via black dot severity, plant growth and yield loss. Pest Technol 4: 45-53.

11. Daami-Remadi M, Zammouri S, El Mahjoub M (2008) Effect of the level of seed tuber infection by Rhizoctonia solani at planting on potato growth and disease severity. Afr J Plant Sci Biotechnol 2: 34-38.
12. Rowe RC, Powelson MR (2002) Potato Early Dying: Management challenges in a changing production environment. Plant Dis 86: 1184-1193.
13. Stevenson WR, Loria R, Franc GD, Weingartner DP (2001) Compendium of Potato Diseases. APS Press, St. Paul, MN, USA,p: 106.
14. Tsror L, Erlich O, Hazanovsky M (1999) Effect of Colletotrichum coccodes on potato yield, tuber quality, and stem colonization during spring and autumn. Plant Dis 83: 561-565.
15. Tsror L, Johnson DA (2000) Colletotrichum coccodes on potato. In: Prusky D, Freeman S, Dickman MB (Eds) Colletotrichum – Host Specificity, Pathology and Host – Pathogen Interaction, APS Press,pp: 362-373.
16. El Bakali AM, Martín MP (2006) Black scurf in potato. Mycologist 20: 130-132.
17. Daami-Remadi M, Zammouri S, El Mahjoub M (2008) Relative susceptibility of nine potato (Solanum tuberosum L.) cultivars to artificial and natural infection by Rhizoctonia solani as measured by stem canker severity, black scurf and plant growth. Afr J Plant Sci Biotechnol 2: 57-66.
18. Daami-Remadi M, Jabnoun-Khiareddine H, Ayed F, El Mahjoub M (2010) Comparative susceptibility of potato cultivars to Verticillium wilt assessed via wilt severity and subsequent yield reduction. Int J Plant Breed 4: 55-62.
19. Davis JR, Huisman OC, Everson DO, Schneider AT (2001) Verticillium wilt of potato: A model of key factors related to disease severity and tuber yield in south-eastern Idaho. Am J Potato Res 78: 291-300.
20. Ayed F, Daami-Remadi M, Jabnoun-Khiareddine H, El Mahjoub M (2006) Effect of potato cultivars on incidence of Fusarium oxysporum f. sp. tuberosi and its transmission to progeny tubers. J Agron 5: 430-434.
21. Daami-Remadi M, Jabnoun-Khiareddine H, Sdiri A, El Mahjoub M (2012) Comparative reaction of potato cultivars to Sclerotium rolfsii assessed by stem rot and tuber decay severity. In Daami-Remadi M. (Ed) Potato Pathology. Pest Technol 6 (Special Issue 2): 54-59.
22. Blok WJ, Lamers JG, Termorshuizen AJ, Bollen GJ (2000) Control of soilborne plant pathogens by incorporating fresh organic amendments followed by tarping. Phytopathology 90: 253-259.
23. Bailey KL, Lazarovits G (2003) Suppressing soil-borne diseases with residue management and organic amendments. Soil Tillage Res 72: 169-180.
24. Janvier C, Villeneuve F, Alabouvette C, Edel-Hermann V, Mateille T, et al. (2007) Soil health through soil disease suppression: Which strategy from descriptors to indicators? Soil Biol Biochem 39: 1-23.
25. Hoitink H, Boehm M (1999) Biocontrol within the Context of Soil Microbial Communities: A Substrate-Dependent Phenomenon.Annu Rev Phytopathol 37: 427-446.

26. Noble R, Coventry R (2005) Suppression of soil-borne plant diseases with composts: a review. Biocontrol Sci Technol 15: 3-20.

27. Bonanomi G, Antignani V, Pane C, Scala F (2007) Suppression of soilborne fungal diseases with organic amendments. J Plant Pathol 89: 311-340.

28. Larkin RP, Honeycutt CW, Olanya OM (2011) Management of Verticillium wilt of potato with disease-suppressive green manures and as affected by previous cropping history. Plant Dis 95: 568-576.

29. Davis JR, Huisman OC, Everson DO, Sorensen LH, Schneider AT (1999) Control of Verticillium wilt of the Russet Burbank potato with corn and barley. Am J Potato Res 76: 367.

30. Wiggins BE, Kinkel LL (2005) Green manures and crop sequences influence potato diseases and pathogen inhibitory activity of indigenous streptomycetes. Phytopathology 95: 178-185.

31. Ochiai N, Powelson ML, Crowe FJ, Dick RP (2008) Green manure effects on soil quality in relation to suppression of Verticillium wilt of potatoes. Biol Fertil Soils 44: 1013-1023.

32. Conn KL, Lazarovits G (1999) Impact of animal manures on Verticillium wilt, potato scab, and soil microbial populations. Can J Plant Pathol 21: 81-92.

33. Tenuta M, Conn KL, Lazarovits G (2002) Volatile Fatty Acids in Liquid Swine Manure Can Kill Microsclerotia of Verticillium dahliae.Phytopathology 92: 548-552.

34. Lu P, Gilardi G, Gullino ML, Garibaldi A (2010) Biofumigation with Brassica plants and its effect on the inoculum potential of Fusarium yellows of Brassica crops. Eur J Plant Pathol 126: 387-402.

35. Manici LM, Caputo F, Babini V (2004) Effect of green manure on Pythium spp. population and microbial communities in intensive cropping systems. Plant Soil 263: 133-142.

36. Campiglia E, Paolini R, Colla G, Mancinelli R (2009) the effects of cover cropping on yield and weed control of potato in a transitional system. Field Crop Res 112: 16-23.

37. Larkin RP, Griffin TS (2007) Control of soilborne diseases of potato using Brassica green manures. Crop Prot 26: 1067-1077.

38. Larkin RP, Griffin TS, Honeycutt CW (2010) Rotation and cover crop effects on soilborne potato diseases, tuber yield, and soil microbial communities. Plant Dis 94: 1491-502.

39. Cherr CM, Scholberg S, McSorley R (2006) Green manure approaches to crop production: A synthesis. Agron J 98: 302-319.

40. White CM, RR Weil (2011) Forage radish cover crops increase soil test phosphorus surrounding radish taproot holes. Soil Fert Plant Nutr 75: 121-130.

41. Charron CS, Sams CE (1999) Inhibition of Pythium ultimum and Rhizoctonia solani by shredded leaves of brassica species. J Am Soc Hortic Sci 124: 462-467.

42. Deacon JW, Mitchell RT (1985) Toxicity of oat roots, oat root extracts, and saponins to zoospores of Pythium spp. and other fungi. Trans Br Mycol Soc 84: 479-487.
43. Mcsorley R, Ozores-Hampton M, Stansly PA, Conner JM (1999) Nematode management, soil fertility, and yield in organic vegetable production. Nematropica 29: 205-213.
44. Melakeberhan H, Mennan S, Ngouajio M, Dudek T (2008) Effect of Meloidogyne hapla on multi-purpose use of oilseed radish (Raphanus sativus). Nematology 10: 375-379.
45. Lampkin NH, Foster C, Padel S, Midmore P (1999) the policy and regulatory environment for organic farming in Europe. Organic farming: Economics and Policy 1 & 2. University of Hohenheim, Stuttgart.
46. Lupwayi NZ, Rice WA, Clayton GW (1998) Soil microbial diversity and community structure under wheat as influenced by tillage and crop rotation. Soil Biol Biochem 30: 1733-1741.
47. Mazzola M, Granatstein DM, Elfving DC, Mullinix K (2001) Suppression of Specific Apple Root Pathogens by Brassica napus Seed Meal Amendment Regardless of Glucosinolate Content.Phytopathology 91: 673-679.
48. Bulluck LR, Ristaino JB (2002) Effect of synthetic and organic soil fertility amendments on southern blight, soil microbial communities, and yield of processing tomatoes.Phytopathology 92: 181-189.
49. Kinkel LL, Stromberg KD, Flor JM, Wiggins E (2001) Green manures influence pathogen inhibitory potential of indigenous antagonist communities in soil. Phytopathology 91: S49.
50. Larkin RP (2013) Green manures and plant disease management. CAB Reviews: Perspectives in Agriculture, Veterinary Science, Nutrition and Natural Resources. CAB Reviews 8-037: 1-10.
51. Neubauer C, Heitmann B, Müller C (2014) Biofumigation potential of Brassicaceae cultivars to Verticillium dahliae. Eur J Plant Pathol 140: 341-352.
52. Zasada IA, Ferris H, Elmore CL, Roncoroni JA, MacDonald JD et al. (2003) Field application of Brassicaceous amendments for control of soilborne pests and pathogens. Online. Plant Health Progress.
53. Hartz TK, Johnstone PR, Miyao EM, Davis RM (2005) Mustard cover crops are ineffective in suppressing soilborne disease or improving processing tomato yield. HortScience 40: 2016-2019.
54. Davis JR, Huisman OC, Westermann DT, Everson DO, Sorensen LH et al. (1996) Effects of green manures on Verticillium wilt of potato. Phytopathology 86: 444-453.
55. Geary B, Ransom C, Brown B, Atkinson D, Hafez S (2008) Weed, disease, and nematode management in onions with biofumigants and metam sodium. HortTechnol 18: 569-574.

56. Molina OI, Tenuta M, El Hadrami A, Buckley K, Cavers C et al. (2014) Potato early dying and yield responses to compost, green manures, seed meal and chemical treatments. Am J Potato Res 91: 414-428.

57. Morra MJ, Kirkegaard JA (2002) Isothiocyanate release from soil-incorporated Brassica tissues. Soil Biol Biochem 34: 1683-1690.

58. Kirkegaard JA, Sarwar M (1998) Biofumigation potential of brassicas I. Variation in glucosinolate profiles of diverse field-grown brassicas. Plant Soil 201: 71-89.

59. Kirkegaard JA, Wong PTW, Desmarchelier JM (1996) In vitro suppression of fungal root pathogens of cereals by Brassica tissues. Plant Pathol 45: 593-603.

60. Sarwar M, Kirkegaard JA (1998) Biofumigation potential of brassicas. II. Effect of environment and ontogeny on glucosinolate production and implications for screening. Plant Soil 201: 91-101.

61. Kruger DHM, Fourie JC, Malan AP (2013) Cover crops with biofumigation properties for the suppression of plant-parasitic nematodes: A review. S Afr J Enol Vitic 34: 287-295.

62. Colla P, Gilardi G, Gullino ML (2012) A review and critical analysis of the European situation of soilborne disease management in the vegetable sector. Phytoparasitica 40: 515- 523.

63. Larkin RP, Honeycutt CW (2006) Effects of different 3-year cropping systems on soil microbial communities and rhizoctonia diseases of potato.Phytopathology 96: 68-79.

64. Lehrsch GA, Gallian JJ (2010) Oilseed radish effects on soil structure and soil water relations. J Sugar Beet Res 47: 1-21.

65. Harding RB, Wicks TJ (2001) In vitro suppression of soil-borne potato pathogens by volatiles released from Brassica residues. In: Porter IJ (ed) Proceedings of the 2nd Australasian soil-borne diseases symposium, Lorne, Australia,pp: 148-149.

66. Hafez SL, Sundararaj P (2004) Biological and chemical management strategies in the sugar beet cyst nematode management. In: Proceedings of the Winter Commodity Schools – 2004, University of Idaho,pp: 243-248.

67. Reddy PP (2013) Biofumigation. In Recent advances in crop protection. Springer,pp: 37-60.

68. Davis JR, Huisman OC, Everson DO, Nolte P, Sorenson LH, et al. (2010) The suppression of Verticillium wilt of potato using corn as a green manure. Am J Potato Res 87: 195-208.

69. Gamliel A, Stapleton JJ (1993) Characterization of antifungal volatile compounds evolved from solarized soil amended with cabbage residues. Phytopathology 83: 899-905.

70. Lawley YE, Teasdale JR, Weil RR (2012) the mechanism for weed suppression by a forage radish cover crop. Agron J 104: 1-10.

71. Pearse IS, Bastow JL, Tsang A (2014) Radish introduction affects soil biota and has a positive impact on the growth of a native plant.Oecologia 174: 471-478.

72. Subbarao KV, Hubbard JC (1996) Interactive effects of broccoli residue and temperature on Verticillium dahliae microsclerotia in soil and on wilt in cauliflower. Phytopathology 86: 1303-1310.

73. Lazarovits G (2010) Managing soilborne disease of potatoes using ecologically based approaches. Am J Potato Res 87: 401-411.

74. Thorup-Kristensen K (2001) Are differences in root growth of nitrogen catch crops important for their ability to reduce soil nitrate-N content, and how can this be measured? Plant Soil 230: 185-195.

75. Nitta T (1991) Diversity of root fungal floras: its implications for soilborne diseases and crop growth. J Agric Res 25: 6-11.

76. Schomberg HH, Endale DM, Calegari A, Peixoto R, MiyazawaM et al. (2006) Influence of cover crops on potential nitrogen availability to succeeding crops in a Southern Piedmont soil. Biol Fertil Soils 42: 299-307.

77. Talgre L, Lauringson E, Makke A, Lauk R (2011) Biomass production and nutrient binding of catch crops. Zemdirbyste-Agriculture 98: 251-258.

78. Lambert DH, ML Powelson, WR Stevenson (2005) Nutritional interactions influencing diseases of potato. Am J Potato Res 82: 309-319.

4

CHAPTER

EFFECT OF NANO-ZINC OXIDE ON THE LEAF PHYSICAL AND NUTRITIONAL QUALITY OF SPINACH

Kisan B, Shruthi H, Sharanagouda H, Revanappa SB and Pramod NK

Department of Biotechnology, College of Agriculture, University of Agricultural Sciences, Raichur-584104, India

ABSTRACT

Spinach (Spinacia oleracea) belongs to family Amaranthaceae and is one of the important and nutritious leafy vegetable consumed in India. The pot culture experiment is carried out during 2014-15 to study the effect of nanozinc oxide particles on the leaf physical and nutritional traits of spinach. The spinach plants were sprayed with graded concentration of zinc oxide nanoparticles (ZnO NPs) after 14 days of sowing. The leaf physical parameters like leaf length, leaf width and leaf surface area are recorded at the time of maturity (45-50 days). The protein, carbohydrate, fat and dietary fiber content in leaf samples are determined. The plants sprayed with ZnO NPs at the concentration of 500 and 1000 ppm showed the increased leaf length, width, surface area and colour of leaf samples when compared to control leaf samples. Similarly treated plants with ZnO NPs at the concentration of 500 and 1000 ppm showed higher values of protein and dietary fibre content in comparison to control leaf samples of spinach. Hence our study suggests that the nano-zinc oxide sprayed spinach is more nutritious to vegetarian diet by providing, protein, fiber and required amount of vegetarian fat to diet.

INTRODUCTION

Spinach (Spinacia oleracea) is a green-leafy vegetable belongs to family Amaranthaceae. It is often recognized as one of the functional foods for its wholesome nutritional,

Citation: Kisan B, Shruthi H, Sharanagouda H, Revanappa SB, Pramod NK (2015) Effect of Nano-Zinc Oxide on the Leaf Physical and Nutritional Quality of Spinach. Agrotechnol 5:135. doi:10.4172/2168- 9881.1000135.

antioxidants and anti-cancer composition. The major micronutrients in spinach are vitamins A (from β-carotene), C, K and folate, and the minerals, calcium, iron and potassium. Spinach also provides fibre and is low in calories. Its tender, crispy, dark-green leaves are one of the favorite ingredients of chefs all around the world. Vegetables are also valuable in maintaining alkaline reserve of the body. They are valued mainly for high carbohydrate, vitamin and mineral contents [1].

Micronutrient fertilizers can increase the tolerance of plants to environmental stresses like drought and salinity [2]. Zinc has been considered as an essential micronutrient for metabolic activities in plants. It regulates the various enzyme activities and required in biochemical reactions leading to formations of chlorophyll and carbohydrates [2,3]. The crop yield and quality of produce can be affected by the deficiency of Zn [4]. Zinc nano-particle is used in various agricultural experiments to understand its effect on growth, germination, and various other properties. Most of the farmers are using either zinc sulfate or EDTA-Zn chelate for soil and foliar applications, however the efficacy is low [5] have demonstrated essentiality and role of zinc in plant growth, reproduction and yield. It has been indicated that the retention time of Zn in the plant system is low and hence, the bioavailability of Zn for long period is not sure with the use of ZnSO4 fertilizer. Under high temperatures conditions ZnSO4 has a large salt index and it may show burning injury if the plants are soft or sensitive [6].

Nano-particles with smaller particle size and large surface area are expected to be the ideal material for use as Zn fertilizer in plants. Application of micronutrient in the form of nano-particles (NPs) is an important route to release required nutrients gradually and in a controlled way, which is essential to mitigate the problems of soil pollution caused by the excess use of chemical fertilizers. A number of researchers have reported the essentiality and role of zinc for plant growth and yield [5,6]. Reynolds [7] demonstrated the use of micronutrients in the form of nano-particles can be used in crop production to increase yield.

It has been postulated that nano-particles are more effective, can be utilized in agriculture for the precision farming and enhance productivity crop yields [7,8]. Several studies are concerned with the synthesis of nanomaterials using biological routes. Only limited studies have been reported on the promotory effects of nanoparticles on plants in low concentrations. Nanoscale titanium dioxide (TiO2) was reported to promote photosynthesis, and growth of spinach. The present study deals with the effect of nano-ZnO, particle suspension as micro-nutrient on the physical and nutritional quality of spinach leaves by foliar spray method. Foliar spray method is more practical from an agronomic standpoint as plant can absorb essential elements through their leaves more efficiently compared to root feeding.

MATERIALS AND METHODS

Zinc oxide nanoparticles of mean size of 50 nm diameter (Sigma Aldrich) were characterized with zeta potential was used in the study. The stock solution of 10,000 ppm solution prepared and dilutions of 0, 100, 500, 1000 ppm were used for the study. The stock solutions were prepared by directly suspending the nano-particles in deionized water and dispersed by ultrasonic vibration (100 W, 40 KHz) for 1 hour. Magnetic bars were placed in the suspensions for stirring before use to avoid aggregation of the particles. The nano-particle solutions used for the experiment purpose were prepared from the stock solutions.

Spinach seeds of variety 'All green variety' were used by sowing in pots (20 cm × 40 cm) filled with equal quantity of soil and watered to field capacity. Proper care was taken to use similar soil in all the pots to minimize soil heterogeneity effects. At 14 days after sowing plants were sprayed with the different concentration of ZnO nano-particles. The leaf samples were harvested at 45th day of sowing and were further processed for proximate analysis.

The physical properties of fresh spinach leafs viz., Length and width were measured using venire callipers and surface area of fresh spinach leafs was determined by using digital Planimeter (Make: Placom; model: KP90N roller-type digital Planimeter). The leaf parameters of fresh spinach leafs i.e., color and water activity were determined using Hunter lab colorimeter (Colour Flex EZ, Hunter Associates LAB INC., C04-1005-631, Taiwan) and Rotronic Hygrolab water activity analyzer (Model: aw-HP23).

The proximate composition of fresh spinach leafs viz. crude protein, crude fat, total ash, crude fiber and carbohydrates were estimated by the recommended methods of the Association of Official Analytical chemists, (2005) in triplicate. Experiments were carried out in triplicate. Data recorded from three replications were subjected to single way analysis of variance (ANOVA) and critical differences were calculated at $p=0.05$ level.

Results and Discussion

Effect of different concentration of ZnO nanoparticles on leaf physical properties such as leaf length, leaf width and leaf surface area etc., were given in Figure 1. The leaf length was maximum in ZnO-NPs treated leaf samples at the concentration of 500 ppm and minimum in control leaf samples. With respect to leaf width, there is no significant variations were observed between treated and control leaf samples of spinach. However, there is slight improvement in leaf width was observed in ZnO-NPs treated leaf samples. The maximum leaf surface area was observed in ZnO-NPs treated leaf samples at concentration of 1000 ppm followed by at 500 ppm and lowest value observed in control leaf samples. Similarly, Prasad [9] reported that, groundnut seeds treated with nanoscale zinc oxide particles with a concentration of 1000 ppm have shown significant increment in germination, shoot length, root length and vigor index over the control samples. Raskar and Laware [8] studied effect of ZnO NPs on seed germination and seedling growth in onion and observed that seed germination increased in lower concentrations of ZnO NPs but showed decrease in values at higher concentrations.

The color values viz., L*, a* and b* of the fresh spinach leaf was found to be is in the range of 37.01-47.11, from -9.79 to -8.97 and 19.68-24.92, respectively and concluded that green in color and water activity of fresh spinach leafs were found to be is in the range of 0.928-0.959. Since the initial moisture content of fresh spinach leaf was more, the presence of available moisture is sufficient to grow micro-organisms also. These values were in confirmation with those reported by Nangula [10] who reported the moisture content of fresh Spinaceaoleraceais found to be 92 g/100 g. Laware and Raskar [6] reported that ZnO-NPs can reduce flowering period by 12-14 days and produce healthy seeds of onion vegetable.

The results of effect of different concentration of ZnO nanoparticles on leaf nutritional traits were given in Table 1.

The Protein content of the raw spinach leaves ranged between 1.54%and 3.99% (Table 1). The plants treated with ZnO-NPs-1000 ppm was found to have the highest protein content (3.99%) and control (1.54%) had the lowest protein content. The protein content to be obtained in the spinach leaf was close to the values previously reported [1] was 2.10 ± 0.15 g. The fat, fiber and carbohydrates content of fresh spinach leaf were found to be is in the range of 0.10-1.57%, 2.01-6.97% and 0.20-0.58%, respectively. Similar results were also concluded by Nangula et al. [10] who reported the fat, fiber and carbohydrates of fresh Spinaceaoleracea was 0.4 g, 3 g and 1 g per 100 g, respectively.

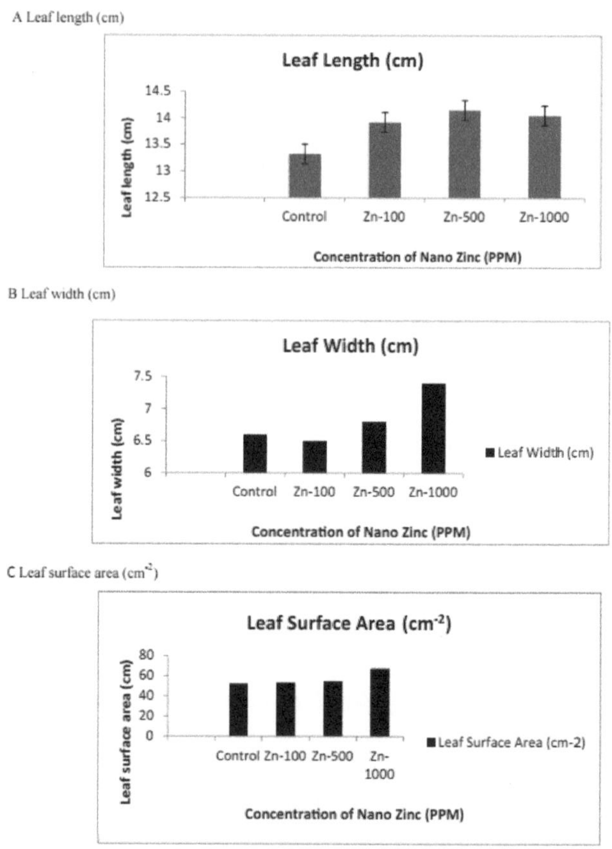

Figure 1: Effect of nano-zinc oxide spray on leaf physical properties of spinach.

Table 1: Effect of nano-zinc oxide spray on leaf nutritional quality of spinach

S. No.	Traits	Control	Zn-100	Zn-500	Zn-1000
1	Protein (%)	1.54	2.69	2.95	3.87
2	Fat (%)	0.1	0.6	1.06	1.07
3	Fiber (%)	2.01	2.23	3.4	6.97
4	Ash (%)	0.18	2.16	1.90	2.06
5	Carbohydrate (%)	0.20	0.23	0.58	0.52

Zinc is an essential micronutrient for normal growth, development, and health of plants and human beings. Zinc enhances cation-exchange capacity of the roots, which in turn enhances absorption of essential nutrients, especially nitrogen which is responsible for higher protein content. Zinc plays vital role in carbohydrate and proteins metabolism as well as it controls plant growth hormone i.e. IAA. Zn is also an essential component of dehydrogenase, proteinase, and peptides enzymes as well as promotes starch formation, seed maturation and production [5]. The application of slow/controlled release fertilizer coated and felted by nano-materials were reported to improve grain yield along with an increase in protein content and a decrease in soluble sugar content in wheat [9].

These results indicated that the nano-zinc oxide enhanced the leaf physical and nutritional properties of spinach leaves. Nano-zinc oxide (1000 ppm) has a potential to be used as a biofortification agent to improve protein and dietary fibre contents of spinach leaves and their by reduce the malnutrition.

REFERENCES

1. RumezaHanif, ZafarIqbal, MudassarIqbal, ShaheenaHanif, MasoomaRasheed (2006) Use of vegetables as nutritional food: role in human health. Journal of Agricultural and Biological Science 1:18-22.
2. Baybordi A (2006) Zinc in soils and crop nutrition. Parivar Press (1stedn)p. 179.
3. Auld DS(2001) Zinc coordination sphere in biochemical zinc sites. Biometals 14:271-313.
4. Jamali G, Enteshari SH, Hosseini SM (2011) Study effect adjustment drought stress application potassium and zinc in corn. Iranian Journal of crop ecophysiology 3:216-222.
5. Fageria NK, Baligar VC, Clark RB (2002) Micronutrients in crop production. Advances in Agronomy 77: 189-272.
6. Laware SL,Raskar SV (2014)Influence of Zinc Oxide Nanoparticles on Growth, Flowering and Seed Productivity in Onion. IntJCurrMicrobiolAppSci3:874-881.
7. Reynolds GH (2002) Forward to the future nanotechnology and regulatory policy. Pacific Research Institute 24:1-23.

8. Raskar SV, Laware SL (2014) Effect of zinc oxide nanoparticles on cytology and seed germination in onion.IntJCurrMicrobiolAppSci3: 467-473.
9. Prasad TNVKV, Sudhakar P, Sreenivasulu Y, Latha P, Munaswamy V, et al. (2012). Effect of nanoscale zinc oxide particles on the germination, growth and yield of peanut.Journal of Plant Nutrition 35: 905-927.
10. Nangula PU, Oelofse A,Duodu KG, Bester MJ, Faber M (2010) Nutritional value of leafy vegetables of sub-Saharan Africa and their potential contribution to human health: A review, Journal of Food Composition and Analysis 23:499-509.

5

CHAPTER

OPTIMUM PATTERN OF COMPOST USED FOR REDUCING ENERGY CONSUMPTION IN MUSHROOM PRODUCTION

Elham Hassanpour[1], Jamal-Ali Olfati[1] and Mohammad Naqashzadegan[2]

[1]Faculty of Agriculture, Horticultural Department, University of Guilan, Rasht, Iran

[2]Faculty of Engineering, Mechanic Department, University of Guilan, Rasht, Iran

ABSTRACT

In mushroom production it is necessary to design a growing pattern to balance yield against cost and to reduce energy use. Effects of use of multiple layer of compost in cultivation on energy consumption in mushroom cropping rooms were examined. Treatments included 1 layer (control), or 2, 3 or 4 layers of compost applied at the mycelium running stage. Numbers of compost layers did not affect fresh weight of mushrooms, number of mushrooms, yield and biological efficiency but hastened pinheads formation. Since there were no differences between layers of compost and control on yield and biological efficiency it appears that 2 layers will be sufficient to improve time to pinhead formation without negative effect on yield.

INTRODUCTION

Agaricus bisporus (Sing) is one of the most important cultivated, edible, mushrooms [1-4]. It contains high amounts of protein, minerals, vitamins D, K, B and sometimes A and C [4,5]. Button mushroom is a natural source of antioxidant agent against the free radicals superoxide radicals (O2), hydroxyl radical (OH), hydrogen peroxide (H2O2) and lipid peroxide radicals and has potential as an anticancer factor [6].

Citation: Hassanpour E, Olfati JA, Naqashzadegan M (2016) Optimum Pattern of Compost used for Reducing Energy Consumption in Mushroom Production. Agrotechnol 5:140. doi:10.4172/2168-9881.1000140.

Mushroom production can help contribute to grow food demand [7]. In Iran there are 1033 units for producing edible mushrooms, of which 704 are used for cultivating button mushroom [8] Iran Statistic Center. Button mushroom has vegetative and reproductive stages. Vegetative growth is on compost, or other substrates [4,9]. In the vegetative stage, temperature must be controlled, with the optimum temperature for mycelium growth being between 25°C and 28°C which require consumption of energy [10]. Agriculture and mushroom production (although it is only a small part of agriculture) is a large consumer of energy [11]. There are times when energy required for agriculture competes for that used by other activities [12]. Energy consumption in mushroom production has increased [13]. Energy consumption in Iran is higher than international standards [14].

It is necessary to design the optimum production pattern for mushrooms to conserve energy [15,16]. Moya et al. believed that energy consumption for heating relates to space among plants; densely spaced plants lead to lower energy consumption. There is no research in relation to reduction of energy consumption in mushroom production. In button mushroom, high yield depends on appropriate conditions of growth and the amount of energy used in cropping rooms needs to be considered. One layer of compost is normally used in button mushroom production. It may be that use of more than one layer of compost will affect growing conditions and consequently yield. This research was conducted to evaluate effects of compost layer number on cultivation on button mushroom development and biological efficiency and effects on energy consumption in spawn running rooms.

MATERIALS AND METHODS

The experiment was conducted with a completely randomized design with 4 treatments and 3 replications in the faculty of Agriculture, University of Guilan, Rasht, Iran (37°16′ N, 51°3′ E). Treatments included 1 (control), or 2, 3 or 4 layers of compost.

Compost blocks (20 kg) were supplied by the Asian Mushroom Company, Karaj, Iran. Block length, width and height were 60, 40 and 20 cm, respectively. Fresh samples of the compost were oven dried at 75°C and dry weight determined. Temperature of compost blocks was measured at 9 AM each day during spawn running for 15 days. After completing spawn running, compost blocks were separated and casing to a 4 cm depth, applied [7] using commercial casing soil (Asian Mushroom Company). After casing, the substrate surface was covered with newspapers and sprayed with water to avoid drying. When the mycelium reached the surface of the casing layer the newspaper was removed and temperature in the room reduced from 25°C to 16°C to provide shocking to induce fruit body formation [17]. Time to pinhead formation was recorded. Mushroom yield was obtained from 2 flushes and mushrooms were harvested daily. During harvesting, numbers and fresh weight of mushrooms were determined. After the final harvest biological efficiency (BE) was determined. Yield and BE were determined according to formulae of Nogueira [18] de Andrade et al. Data were subjected to ANOVA in SAS (ver. 9.1, SAS Institute, Inc., Cary, N.C.). Means were separated with using Turkey's test.

Results

Changes of temperature during spawn running in compost treatments were similar (Figure 1). Temperature increased from day 1 to day 7 for all layers of compost. Temperature decreased, finally reaching 21°C and 23°C for 1 layer and 4 layers of compost, respectively; other treatments were intermediate. Increased temperature in first week may be due to reduced ventilation.

Numbers (average 163.72), fresh weight (average 1828.38 g/ tray), and BE of mushrooms (29.79%) and yield (average 6.23) were not affected by treatment. Days to pinhead formation was affected by treatment (Table 1). One layer of compost (control) needed the most time for pinhead formation, 16.33 (average) days. The least time for pinhead formation was for the 3 layer treatment, 11.5 days. The multilayer treatments were similar to each other and less than for the control (Table 2).

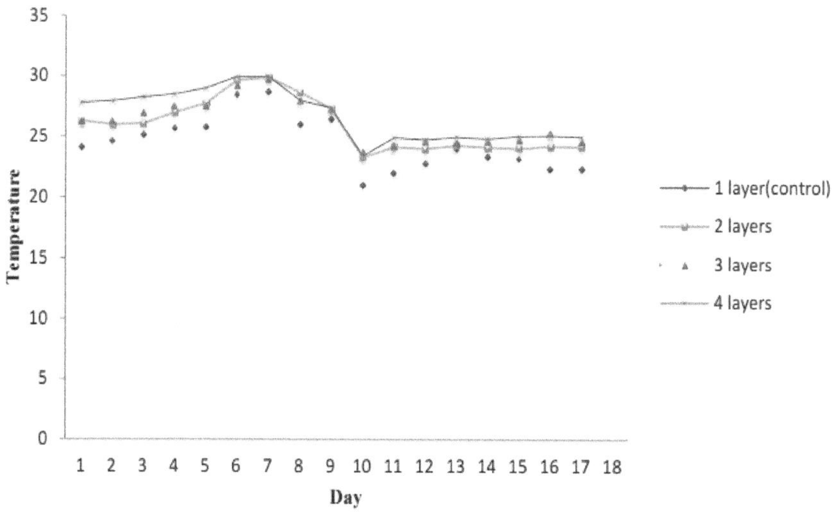

Figure 1: Temperature change during spawn running in treatments.

Table 1: Effect of multi-layer cultivation on fresh weight and number of mushrooms in first, second and total flushes

		Mean squares					
Source of variation	df	Fresh weight			Number of mushroom		
		Flush 1	Flush 2	total	Flush 1	Flush 2	total
Compost layer	3	164908.53 ns	1.88 ns	72692.22 ns	0.41 ns	2.81 ns	1591.16 ns
Error	8	66357.31	2.55	127106.7	0.54	1.49	831.96
CV (%)		18.6	20.94	18.54	18.92	9.55	17.44

ns = non-significant at $P < 0.05$.

Table 2: Effect of multi-layer cultivation on biological efficiency (BE %), yield (Y %) and days to pinhead formation in first, second and total flushes

		Mean squares					
Source of variation	df	B.E. (%) Flush1	B.E. (%) Flush2	B.E. (%) Total	Y (%) Flush1	Y (%) Flush2	Days to pinhead formation
Compost layer	3	45.45 ns	0.19 ns	19.45 ns	4.10 ns	0.07 ns	13.92**
Error	8	17.59	0.32	33.63	1.66	0.07	0.38
CV (%)		18.51	10.16	18.51	18.61	19.19	4.69

ns, **, non-significant and significant at $P < 0.01$.

Discussion

Growing mycelium produces heat and temperature increases during the first days of growth and then decreases [19]. Increasing layers of compost can increase temperature may be due to reduce ventilation and CO2 exchange in running room. There was likely insufficient air between compost blocks so heat and CO2 was retained [20]. Several layers of compost increases temperature and CO2 stimulating mycelial growth during the mycelium running stage [10]. In other hand a positive correlation between the thermophiles population and amount of air supplied was recorded in previous research [21]. The reason for this needs additional study.

Optimum temperature for mycelium growth is 24°C-25°C and the maximum temperature is 28°C [7]. The temperature in the 3-4 layer treatment was at, or below this temperature. Appropriate ventilation results in moderate environment and CO2 content stability in fruiting stage [10,19,20]. In compact growth substrate there is low CO2, and respiration and growth decreases. Increased temperature increases evaporation and transpiration which results in decreased mycelial growth [20]. Growth and development of pathogens could increase because of anaerobic conditions [10].

Imbalanced mycelial running in substrate and weak, or inadequate, mycelial growth is the consequences of less than optimum compost temperature [7]. Temperature increasing to 30°C causes mycelial death and increased activity of microorganisms and insects Megaselia nigra M. and Lycoriella auripila W. feeding on mycelia [10]. However some authors Vedder and Hussain et al., [22,23] believe that for better mycelial growth average compost temperature should be 30°C and mycelium only die after a prolonged exposure to above 34°C.

Several layers of compost resulted in earlier pinhead appearance [24]. Evaluated this characteristic in different casing soil with maximum and minimum days to pinheads appearance being 23.75 and 12.25, respectively. Our results (maximum 16.33 days, minimum 11.5 days) were lower than theirs indicating multilayer cultivation reduced time for pinhead formation. Use of 2 layers of compost could be beneficial in mushroom cultivation.

CONCLUSION

Multilayer cultivation may result in consuming less energy for heating the cropping room without reduction in yield. In multilayer cultivation as a new idea on mushroom cultivation we put more compost in a mycelium running room and divide ready to fruiting compost in fruiting room. So only a translocation add to our cost while energy consumption for mycelium running room or cooling for shocking specially in warm season at least two fold decreased.

REFERENCES

1. Gbolagade J, Ajayi A, Oku I, Wankasi D (2006) Nutritive value of common wild edible mushrooms from southern Nigeria. Global Journal Biotechnology Biochemistry 1: 16-21
2. Toker H, Baysal E, Yigitbasi ON, Colak M, Peker H, et al. (2007) Cultivation of Agaricusbisporus on wheat straw and waste tea leaves based composts using poplar leaves as activator material. African Journal of Biotechnology 6: 204-212.
3. Mehta BK, Jain SK, Sharma GP, Doshi A, Jain HK (2011) Cultivation of button mushroom and its processing: An techno-economic feasibility. International Journal of Advanced Biotechnology and Research 2: 201-207
4. Ebadi A, Alikhani HA, Rashtbari M (2012) Effect of plant growth promoting bacteria (PGPR) on the morphophysiological properties of Button mushroom Agaricusbisporus in two different culturing beds. International Research Journal of Applied and Basic Sciences 3: 203-212.
5. Saiqa S, Nawaz BH, Asif HM (2008) Studies on chemical composition and nutritiveevaluation of wild edible mushrooms. Iran Journal Chemichal Engineering 27: 151-156.
6. Abah SE, Abah G(2010) Antimicrobial and antioxidant potentials of Agaricusbisporus.Advances in Biological Research4: 277-282.
7. Farsi M, Pouriyanfar HR (2011) Cultivation and breeding of the white button mushroom. Mashhad Jihad University Press, Mashhad, Iran.
8. Iran Statistic Center (2012)Statistical pocketbook of the Islamic republic of Iran
9. Noble R, Dobrovin-Pennington A, Hobbs P, Rodger A, Pederby J (2009) Volatile C8 compounds and pseudomonas influence primordium formation of Agaricusbisporus.Mycologia 101: 583-591.
10. Jafarniya S, Khosroshahi M, Karami SM (2010) Mushroom cultivation: appropriate technology for mushroom growers. Misagh Press, Tehran, Iran.
11. Alam MS, Alam MR, Islam KK(2005) Energy flow in agriculture: Bangladesh. AmericanJournal Environmental Science 1: 213-220.
12. Pahlavan R, Omod M, Akram A (2012) Application of data envelopment analysis forperformance assessment and energy efficiency improvement opportunities in greenhouse cucumber production. Journal of Agricultural Science and Technology 14: 1465-1475.
13. Mehrabi H, Esmaili A (2011) Analysis energy input-output in Agriculture in Iran. Agriculture Economic and Development Journal 74: 12-18
14. Kazemi A, Mehregan MR, Shakouri H, Hosseinzadeh M (2012) Energy resource allocation in Iran: A fuzzy multi-objective analysis. Procedia-Social and Behavioral Sciences 41: 334-341.
15. Faizi F, Nourani M, Ghaedi AK, Mahdavinejad MJ (2011) Design an optimum pattern of orientation in residential complexes by analyzing the level of energy consumption. Proceeding Engineering 21: 1179-1187.

16. Moya A, Mehlitz T, Yildiz I, Kelly SF, Hardin C (2008) Simulated effects of dynamic row spacing on energy and water conservation in semi-arid central California greenhouses. Department of bio-resource and agricultural and mechanical engineering. Polytechnic State University, Marietta, CA.

17. Femor TR, Randle PE, Smith JF (1985) Compost as a substrate and its preparation. In: FLEGG PB, SPENCER DM, WOO DA: The biology and technology of the cultivated mushroom. Wiley, Chichester, UK, pp:81-109.

18. De Andrade M, Kopytowski J, Minhoni M, Coutinho L, Figueiredo M (2007) Productivity, biological efficiency and number of Agaricusblazeimushrooms grown in compost in the presence of Trichoderma sp. and Chaetominumolivacearumcontaminants.Brazilian Journal of Microbiology 38: 243-247.

19. Kashi A (2005) Edible mushroom cultivation. Agricultural Education Press, Karaj, Iran.

20. Peyvast GH, Olfati JA (2005) Improved cultivation of edible mushrooms. DaneshpazirPress,Tehran, Iran.

21. Wakchaure GC, Meena KK, ChoudharyRl, Singh M, Yandigeri MS (2013) Animprovedrapid composting procedure enhance the substrate quality and yield of Agaricusbisporus. African Journal of Agricultural Research 8: 4523-4536.

22. Hussain S, Ali MA, Ahsan A, Ali H, Siddique M (2004) Temperature requirement of button mushroom Agaricusbitorquis for mycelia growth on three different composts. Journal of agricultural research 42:3-4.

23. Yilmaz F, Baysal E, Toker H, Colak M, Yigitbasi O, Simsek H (2007) An investigationon pinhead formation time of Agaricusbisporus on wheat straw and wastetea leaves based composts using some locally available peat materials and secondarycasing materials. African Journal of Biotechnology 6: 1655-1664.

6
CHAPTER

EFFECT OF FOLIAR APPLICATION WITH ALGAE AND PLANT EXTRACTS ON GROWTH, YIELD AND FRUIT QUALITY OF FRUITFUL MANGO TREES CV. FAGRI KALAN

El-Sharony TF[1], El-Gioushy SF[2] and Amin OA[1]

[1]Horticultural Crops Technology Department, National Research Centre, Dokki, Giza, Egypt

[2]Department of Horticulture, Faculty of Agriculture, Benha University, Moshtohor, 13736 Qaluiobia, Egypt

ABSTRACT

An experiment was conducted during two successive seasons 2013 and 2014 on mango trees cv. Fagri Kalan to study the effect of foliar spray with some extracts (algae at 2%, roselle at 10% and garlic at 5%) applied four times along each season (at full bloom, fruit set, one and two months later from fruit set) either each extract was sprayed solely or in different combinations each other. The influence was evaluated through the response of leaf N, P, K content, vegetative growth, yield and some fruit physical and chemical characteristics of mango cv. Fagri Kalan. The obtained results showed that, spaying algae extract at 2% either alone or combined with one of two plants extracts (10% roselle and 5% garlic) particularly both plant extracts together surpassed in most cases other investigated spray treatments for improving most measurements of fruiting (fruit set, retention, yield and fruit quality), vegetative growth, and nutritional status, (leaf N, P and K content). Besides, roselle and garlic extracts spray had also a noticeable positive effect particularly on some fruit physical and chemical characteristics. Therefore, algae and investigated two plants extracts could be safely recommended as a natural biostimulants application for improving most desirable measurements of mango trees cv. Fagri Kalan grown under the same experimental condition.

Citation: El-Sharony TF, El-Gioushy SF, Amin OA (2015) Effect of Foliar Application with Algae and Plant Extracts on Growth, Yield and Fruit Quality of Fruitful Mango Trees Cv. Fagri Kalan. J Horticulture 2:162. doi:10.4172/2376-0354.1000162.

INTRODUCTION

Mango (Mangifera indica L.) belongs to the family Anacardiaceae considered as one of the most important fruits of the tropical and subtropical countries of the world. It grows under a wide range of climatic and soil conditions. In Egypt, mango ranks the third after citrus and grapes, whereas its total area of fruitful orchards reached approximately 101303 ha producing about 712537 tons annually [1]. However, lower yield with poor quality is one of the main problems facing mango growers in the new reclaimed lands particularly under sandy soil conditions. This may be attributed to the relative lower and/or the complete lacking of such nutrient and essential elements. Meanwhile, various trials were done to raise the productivity and fruit quality of much fruit species by the wasteful use of chemical fertilizers especially mineral ones. For avoiding the excessive use of the mineral chemical fertilizers and their harmful effect particularly on human health consequently using of some natural biostimulants such as plants and algae extracts that characterized by their richer contents with nutrient elements, growth regulators, antioxidants and vitamins were investigated for being applied as an effective nutritive application that could be replace partially or completely the aforesaid harmful ones.

Algae extract as a new biofertilizer containing some macro and micro nutrients i.e. (N, P, K, Ca, Mg and S) and (Zn, Fe, Mn, Cu, Mo and Co) as well as some growth regulators, polyamines and vitamins required to be applied for improve nutritional status, vegetative growth, yield and fruit quality in different fruits orchards [2-4].

Roselle (Hibischus sabdariffa L.) belongs to the family Malvaceae. The edible portion is not flowers but calyces. Roselle main ingredients are vitamins C, A, D, B1 and B2, antioxidants, anthocyanins, Fe, Mg and omega3 beta-carotene (Bruneton, 2001). Furthermore, using extracts of roselle improve the nutritional status and yield as well as some physical and chemical characteristics of Valencia orange fruits. The beneficial effects of roselle extracts on building plant pigments and organic foods surely reflected on advancing maturity and improving fruit quality [5].

Garlic (Allium sativum) extract contains enzymes and more than 200 chemical compounds, some of its volatiles are more important i.e. Allicin that gives garlic its antibiotic properties. Its higher contents of volatile and sulphur compounds put both in the top due to their real and essential roles they play in fruiting process of various fruit crops [6-8]. Garlic also contains vitamins, minerals, flavonoids, ascorbic acid, sulphur and trace of iodine. Seventeen amino acids are found in garlic, including eight essential ones. Effect of garlic extract on plant characters has been interpreted by Sayeeda and Ahmed [9] and El- Shayeb [10]. They reported that garlic extract showed comparatively greater efficacy on promoting growth and nutrition status of groundnut. Abbas et al. [11] also indicated that spray of garlic extract on cucumber at (1:1) increased auxins and decreased gibberellins-like substances, whereas garlic extract at (0.5:1) decreased auxin contents.

Therefore, this study aimed to improve growth and productivity of "Fagri Kalan" mango trees by applying some natural extracts i.e. algae, roselle and garlic sprayed either solely or combined each other.

MATERIALS AND METHODS

This study was carried out during two successive 2013 and 2014 experimental seasons on fruitful mango trees cv. Fagri Kalan grown in Research and production Station orchard at El-Nobaria region, Behera Governorate, Egypt. Twenty four Fagri Kalan mango trees of ten years old, planted at 3 × 5 m apart in a sandy soil under drip irrigation system were carfuly selected as being healthy, disease free and uniform as possible in their vigour and size to study the effect of foliar spray with some plants and algae extracts. All devoted trees received regularly the same horticultural practices (fertilization, irrigation, pest control, pruning etc.,) adopted in mango orchards of the region. The complete randomized block design with three replications (each replicate was represented by a single tree) was used for arranging the following eight spray treatments:

- Control (spraying with water only)
- Algae extract at 2%

Roselle (Hibischus sabdariffa L.) extract at 10% (dried hibiscus leaves brought from the local market, then grinded and taking soft powder to prepare the desired concentration by dissolving 250 g of fine powder in 250 ml of distilled water in an electric mixer and mixing the mixture for 15 m, then let the mixture for 30 m and filtering several times and kept at low temperature (4°C) till used.

Garlic (Allium sativum) extract at 5% (250 g of garlic cloves were mixed with 250 ml tap water and thawed three times. Water was added to a final volume of 1 liter before filtering. Where the filtrate was adjusted to 1: l by volume before being used).

- Roselle extract at 10%+garlic extract at 5%
- Roselle extract at 10%+algae extract at 2%
- Garlic extract at 5%+algae extract at 2%
- Roselle extract at 10%+garlic extract at 5%+algae extract at 2%
- Triton B at 0.1% as a wetting agent was used with each treatment even control.

Taking into consideration that all investigated spray treatments were applied four times along each growing season i.e., just at full bloom, fruit set, one and two months after fruit set had been taken place prior to harvesting.

Measurements and Analysis

The effect of the differential investigated spray treatments was evaluated through the response of the following measurements:

Vegetative growth parameters: On each tree four main branches (limbs/scaffolds) similar in their vigour and well distributed around its periphery (each towards one geographic direction) were carefully selected and labeled. Then, twelve spring cycle shoots (three per every labeled limb were randomly selected) for determining the following growth measurements:

- Number of new shoots
- Shoot length (cm)

- Shoot diameter (cm)
- Numbers of leaves\shoot

Leaf area in (cm2) according to Ahmed and Morsy [12] using the following equation: Leaf area (cm2)=0.70 (leaf length x leaf width)-1.06.

Leaf N, P and K contents as percentages in dry matter of leaves samples at 2nd week of July during both seasons were determined as follows:

- Nitrogen, after Pregel [13] (using Micro-Kjeldahl).
- Phosphorus, according to method described by Champman and Pratt [14].
- Potassium, according to Brown and Lilleland [15].
- Average number of fruit set/panicle was recorded.

Fruit retention percentage was recorded at mature stage (a week before harvest).

Tree yield was estimated as weight of harvested fruits (Kg) per tree on the last week of July in each season.

Fruit quality:

- Five mature fruits were randomly sampled from those harvested ones per each tree for determining the following fruit physical and chemical properties:
- Fruit physical characteristics: such as average fresh weights of the whole fruit, pulp, peel and stone (seed) in grams were determined. Besides, pulp/peel ratio was also estimated.

Fruit chemical characteristics: samples of fruit juice were used to determine the total soluble solids percentage (TSS%) using hand refractometer and total sugars as (g/100 g fresh weight) after method described by Dubois et al. [16]. Moreover, total acidity as percentage of citric and malic acids was determined according to A.O.A.C [17]. Fruit ascorbic acid (V.C) content as milligrams ascorbic acid /100 ml juice, was determined according to A.O.A.C [17].

Statistical Analysis

The obtained data was subjected to analysis of variances (ANOVA) according to Snedecor and Cochran [18]. Mstat-C program was used to calculate least significant differences LSD letters were used for comparing between means of different treatments according to Wallar and Duncan [19] at probability of 0.05 level [20].

Results and Discussions

In this regard foliar spray with some algae and plant extracts were investigated regarding their effects on some measurements of vegetative, growth, nutritional status, productivity and fruit quality of fruitful mango trees cultivar Fagri Kalan during 2013 & 2014 seasons.

Effect on Vegetative Growth Parameters

Data obtained during both seasons as shown in Table 1 displayed that all/most spray treatments with algae and plant extracts resulted in an increase over control for the differential investigated growth parameters (No. of new developed shoots, shoot length, shoot thickness, No. of leaves/shoot and average leaf area). The increase in all evaluated growth parameters exhibited by the differential investigated spray application with algae and plant extracts over control (tap water spray) varied not only from one spray treatment to another but also, each evaluated growth easements reflect its own rate of response. Anyhow, it could be obviously concluded that the response was more pronounced and increases over control were significant as the three spray treatments (roselle 10%+algae extract 2%, garlic 5%+algae extract 2% and roselle 10%+garlic 5% +algae extract 2%). Such trend was true during both seasons with few exceptions particularly in 2nd season for the number of new developed shoots, when differences were completely absent.

Table 1: Effect of foliar application with algae and plant extracts on some vegetative growth parameters of mango trees cv. Fagri kalan during 2013 and 2014 seasons. Means within each column followed by different letter/s are statistically different at 5% level.

Treatments	No. New shoots		Shoot length (cm)		Shoot thickness (mm)		No. Leaves/ shoot		Leaf area (cm^2)	
	2013	2014	2013	2014	2013	2014	2013	2014	2013	2014
T1: Control	3.00 b	3.50 a	18.00 d	16.00 b	6.00 b	7.50 c	12.00 c	15.00 b	28.46 d	47.07 b
T2: Algae Extract 2%	3.50 ab	3.50 a	26.00 bcd	23.00 b	7.50 ab	9.50 a	17.50 b	19.50 a	59.67 ab	50.65 b
T3: Roselle Extract 10%	3.00 b	3.00 a	20.50 d	19.50 b	6.00 b	7.50 c	14.50 bc	16.00 b	41.62 c	49.59 b
T4: Garlic Extract 5%	3.50 ab	3.00 a	21.00 d	20.50 b	7.00 ab	7.50 c	15.50 bc	17.00 b	47.44 bc	45.32 b
T5: Roselle 10% + Garlic Extract 5%	3.50 ab	3.00 a	24.50 cd	22.00 b	7.50 ab	8.00 bc	16.35 bc	17.00 b	56.00 ab	50.65 b
T6: Roselle 10% + Algae Extract 2%	3.67 ab	3.50 a	32.00 abc	31.50 a	8.00 a	9.50 a	18.50 b	25.00 a	61.50 a	51.51 b
T7: Garlic 5% + Algae Extract 2%	4.00 ab	4.00 a	35.00 ab	33.00 a	9.00 a	9.50 a	22.50 a	25.00 a	61.94 a	62.35 a
T8: Roselle 10% + Garlic 5% + Algae Extract 2%	5.00 a	4.00 a	41.00 a	34.00 a	9.50 a	10.00 a	24.50 a	26.00 a	65.56 a	64.94 a

On the other hand, the least increase over control in all evaluated growth measurements was always in concomitant to the Fagri Kalan mango trees sprayed with roselle extract at 10% and garlic extract at 5% either each was applied solely or both combined each other i.e. roselle extract 10%, garlic extract 5% and roselle 10% +garlic extract 5%, where the increase was not significant and it could be safely neglected from the statistic point during both seasons. In addition, the effect of spray treatment with algae extract at 2% solely was in between aforesaid two extremes.

Effect on Nutritional Status (leaf N, P, K Contents)

Table 2 reveals that the response of leaf N, P, and K contents of Fagri Kalan mango trees to the investigated foliar spray with some extracts of algae, garlic and roselle was to some extent similar to that previously detected with growth measurements, pertaining the relative (slight/ clear) increase over control.

Table 2: Effect of foliar application with algae and plant extracts on leaf mineral content (NPK) percentages of mango trees cv. Fagri kalan during 2013 and 2014 seasons. Means within each column followed by different letter/s are statistically different at 5% level

Treatments	N (%)		P (%)		K (%)	
	2013	2014	2013	2014	2013	2014
T1: Control	1.07 c	1.12 b	0.21b	0.22 c	0.88 b	0.89 d
T2: Algae Extract 2%	1.15 ab	1.18 ab	0.25 ab	0.27 ab	0.95 ab	0.97 abc
T3: Roselle Extract 10%	1.09 bc	1.12 b	0.29 a	0.29 a	0.89 ab	0.91 cd
T4: Garlic Extract 5%	1.13 abc	1.14 b	0.25 ab	0.28 ab	0.91 ab	0.91 cd
T5: Roselle 10% + Garlic Extract 5%	1.14 abc	1.17 ab	0.25 ab	0.30 a	0.91 ab	0.94 bcd
T6: Roselle 10% + Algae Extract 2%	1.17 ab	1.24 a	0.22 ab	0.25 bc	0.95 ab	0.99 ab
T7: Garlic 5% + Algae Extract 2%	1.18 a	1.24 a	0.24 ab	0.25 bc	0.96 ab	1.01 ab
T8: Roselle 10% + Garlic 5% + Algae Extract 2%	1.19 a	1.25 a	0.25 ab	0.30 a	0.98 a	1.02 a

On the other hand, the trend of response for both leaf nitrogen and potassium contents followed typically the same one resulted in growth parameters particularly the superiority of (roselle 10% +algae extract 2%, garlic 5% +algae extract 2% and roselle 10% +garlic 5% +algae extract 2%) discerningly followed by (algae at 2%), (garlic at 5% solely and/or garlic 5% +roselle 10%) and (roselle 10% solely). However, leaf phosphorus content, the trend of response was slightly modified. Herein, the differences were in most cases not significant and foliar spray with roselle extract at 10% either solely or combined to (garlic 5%) and (garlic 5% +algae 2%) induced generally the riche of Fagri Kalan mango leaves in their phosphorus contents.

Effect on some Productivity Measurements

In this respect fruit set, fruit retention and yield/tree in Kg were the investigated measurements of mango trees productivity.

Table 3: Effect of foliar application with algae and plant extracts on fruit setting, fruit drop, fruit retention and yield of mango trees cv. Fagri klan during 2013 and 2014 seasons. Means within each column followed by different letter/s are statistically different at 5% level

Treatments	Fruit set/panicle		Fruit retention (%)		No. of fruits/tree		Yield (kg/tree)	
	2013	2014	2013	2014	2013	2014	2013	2014
T1: Control	5.92 e	6.08 d	0.60 f	0.65f	24.00 f	28.67 g	11.59 d	12.27 ef
T2: Algae Extract 2%	7.98 cd	8.07 bc	1.12 d	1.24 c	38.67 c	43.33 cd	16.65 c	19.00 cd
T3: Roselle Extract 10%	6.23 e	6.39 d	0.75 e	0.94 e	31.33 e	37.00 ef	12.74 d	14.95 e
T4: Garlic Extract 5%	7.17 de	7.34 cd	0.85 e	1.02 de	31.00 e	35.67 f	12.75 d	15.57 e
T5: Roselle 10% + Garlic Extract 5%	7.69 cd	7.74 c	1.04 d	1.20 cd	35.67 d	40.33 de	15.28 c	17.94 de
T6: Roselle 10% + Algae Extract 2%	9.01 bc	9.20 ab	1.28 c	1.64 b	44.00 b	47.33 b	19.93 b	21.54 bc
T7: Garlic 5% + Algae Extract 2%	9.69 ab	9.76 a	1.46 b	1.70 ab	46.00 ab	51.67 b	21.71 ab	24.77 ab
T8: Roselle 10% + Garlic 5% + Algae Extract 2%	10.75 a	10.43 a	1.71 a	1.86 a	47.67 a	57.00 a	23.33 a	28.33 a

It's quite evident as shown from tabulated data in Table 3 that, three measurements followed to great extent the same trend pertaining their response to the differential investigated spray treatments with some algae and plant extracts. Herein, the seven spray treatments with algae, garlic and roselle extracts (either each was applied solely or in combination each other/s) resulted significantly in increasing three productivity parameters over control (water spray), during two seasons of study. Meanwhile, (roselle 10% +algae extract 2%, garlic 5% +algae extract 2% and roselle 10% +garlic 5% +algae extract 2%) spray treatments were statistically the superior with comparison to four other extracts sprayed treatments regarding their increasing effect over control during two seasons of study. However, foliar spray with three extracts of algae, garlic and roselle together i.e., roselle 10% +garlic 5% +algae extract 2% treatment surpassed in most cases two other effective ones (roselle 10% + algae extract 2% and garlic 5% +algae extract 2% treatments) particularly roselle 10% +algae extract 2% ones during both 2013 and 2014 seasons. Such trend of the beneficial effect of different spray treatments with investigated three extracts of algae, garlic and roselle on three productivity measurements of Fagri Kalan mango trees may be logically explained as a real reflection of the enhancement resulted by such extracts in both growth and nutritional status measurements.

Effect on Fruit Quality

Fruit physical characteristics: In this concern, average weight in g. of the (whole fruit, fruit pulp, peel, stone and the pulp/fruit ratio) were the investigated of the fruit physical characteristics of Fagri Kalan mango cv.

Data obtained during both 2013 and 2014 seasons are presented in Table 4 displayed that all/most spray treatments with algae and plant extracts resulted in an increase over control for the differential investigated fruit physical characteristics (average weight of whole fruit, fruit pulp, peel, stone and the pulp/fruit ratio). Anyhow, it could be obviously concluded that the response was more pronounced and increases over control were significant as the three spray treatments (roselle 10% +algae extract 2%, garlic 5% +algae extract 2% and roselle 10% +garlic 5% +algae extract 2%). Such trend was true during both seasons 2013 & 2014.

On the other hand, the least increase over control in all evaluated fruit physical characteristics was always in concomitant to the Fagri Kalan mango sprayed with roselle extract at 10% and garlic extract at 5% either each was applied solely or both combined each other i.e. roselle extract 10%, garlic extract 5% and roselle 10% +garlic extract 5%.

Fruit chemical properties: Total soluble solids (TSS%), total sugar%, total acidity and vitamin C (ascorbic acid) were the investigated chemical properties in response to foliar spray treatments with algae, garlic and roselle extracts. Data obtained during two seasons are presented in Table 5 illustrated the effect of algae and plant extracts as foliar spray on fruit chemical properties.

Table 4: Effect of foliar application with algae and plant extracts on some fruit physical properties of mango cv. Fagri kalan during 2013 and 2014 seasons. Means within each column followed by different letter/s are statistically different at 5% level

	Fruit weight (g)		Pulp weight (g)		Pell weight (g)		Seed weight (g)		Pulp/Fruit ratio	
Treatments	2013	2014	2013	2014	2013	2014	2013	2014	2013	2014
T1: Control	399.67 e	393.33 d	301.99 e	292.97 f	58.48 h	59.97f	39.20 g	40.39 g	75.56 a	74.48 a
T2: Algae Extract 2%	430.67 d	445.00 c	322.46 d	333.72 cd	64.65d	65.65 d	43.59 d	45.63 d	74.87 a	74.99 a
T3: Roselle Extract 10%	406.33 e	404.33 d	304.58 e	299.58 e	60.78 g	63.17 e	40.97 f	41.58 f	74.96 a	74.09 a
T4: Garlic Extract 5%	411.33 e	437.67 c	307.83 e	329.56 d	61.60 f	63.49 e	41.90 ef	44.62 e	74.84 a	75.30 a
T5: Roselle 10% + Garlic Extract 5%	428.33 d	438.00 c	322.66 d	327.70 d	62.92 e	65.03 d	42.75 de	45.27 de	75.33 a	74.82 a
T6: Roselle 10% + Algae Extract 2%	452.67 c	455.00 bc	338.29 c	336.34 c	68.12 c	70.15 c	46.26 c	48.51 c	74.73 a	73.92 a
T7: Garlic 5% + Algae Extract 2%	472.00 b	479.33 ab	353.95 b	357.86 b	70.01 b	71.49 b	48.04 b	49.98 b	74.99 a	74.66 a
T8: Roselle 10% + Garlic 5% + Algae Extract 2%	489.33 a	497.00 a	364.23 a	367.60 a	74.77a	76.37 a	50.33 a	53.03 a	74.43 a	73.96 a

It is quite clear that, spraying algae and plants extracts had an obvious effect on enhance the fruit chemical properties. Whereas, 10% roselle+5% garlic+2% algae extracts showed to be the most effective treatment for inducing the highest values of total soluble solids percentage, total sugars and ascorbic acid content followed in a descending order by 5% garlic+2% algae extracts and 10% roselle+2% algae extracts in both seasons when compared with control which recorded the lowest values of these concerns in both seasons. However, other treatments came in between in both seasons of the study. While, there were no significant effects of different treatments on fruit content of total acidity in both seasons.

Table 5: Effect of foliar application with algae and plant extracts on some fruit chemical properties of mango cv. Fagri kalan during 2013 and 2014 seasons. Means within each column followed by different letter/s are statistically different at 5% level.

	Total soluble solids TSS (%)		Total Sugar (g/100 g FW)		Acidity (%)		Ascorbic acid (mg/100 ml juice)	
Treatments	2013	2014	2013	2014	2013	2014	2013	2014
T1: Control	15.00 b	12.77 c	10.53 a	9.94 c	0.50 a	0.61 a	32.57 a	31.33 cd
T2: Algae Extract 2%	18.20 ab	16.13 ab	12.74 a	11.34 ab	0.36 a	0.51 a	37.72 a	32.77 bc
T3: Roselle Extract 10%	15.93 ab	14.53 bc	11.17 a	10.16 bc	0.47 a	0.61 a	32.65 a	31.88 bc
T4: Garlic Extract 5%	16.07 ab	15.20 abc	11.31 a	10.64 abc	0.47 a	0.58 a	34.33 a	32.28 bc
T5: Roselle 10% + Garlic Extract 5%	16.40 ab	15.47 abc	11.61 a	10.80 ab	0.44 a	0.53 a	37.47 a	32.50 bc
T6: Roselle 10% + Algae Extract 2%	18.47 ab	16.40 ab	12.96 a	11.50 ab	0.45 a	0.54 a	37.89 a	34.59 ab
T7: Garlic 5% + Algae Extract 2%	19.47 a	17.27 ab	13.70 a	12.06 a	0.43 a	0.57 a	39.85 a	37.07 a
T8: Roselle 10% + Garlic 5% + Algae Extract 2%	19.87 a	17.73 a	13.97 a	12.45 a	0.44 a	0.54 a	40.52 a	37.15 a

DISCUSSIONS

The outstanding effects of algae and plant extracts on vegetative growth, leaves content of N, P and k, fruit set, fruit retention, yield as well as fruit quality of mango cv. Fagri

Kalan might be due to the content of these extracts on nutrients such as (N, P, K, Ca, Mg, and S), vitamins, some growth regulators and polyamines as algae extract. Also, may be due to these extracts contents of antioxidants, vitamins C, anthocyanins and carotene as roselle extract or may be due to these extracts contain of enzymes, volatile compounds and sulphur compounds as garlic extract. The use of such materials on the plant in general and especially on mango trees will improve the physiological and nutritional status of the trees and increase the yield and fruit quality.

These results are in harmony with those obtained by Chawdhury et al. [21], Karim and Rahim [22], Abd El-Motty et al. [23], Ahmed et al. [24] and Adel Abed El Hamied [25] on mango. They reported that, spraying mango trees with algae extracts alone or combined with natural or plant extracts was very effective in improving fruit set, fruit retention, yield and enhanced fruit quality. Also, Ahmed et al. [5] and Hegab et al. [26] on citrus, Hafez et al. [27] on olive and Hanafy et al. [28] on plant extract.

CONCLUSION

From the above results it could be concluded that, using algae extract at 2% alone or in combination with 10% roselle or 5% garlic extracts had a clear effects on vegetative growth, leaves contents of N, P and K, fruit set, fruit retention, yield and fruit quality of mango cv. Fagri Kalan. It was also noted that spraying extracts of roselle at 10 % and garlic at 5% had a good effect on some fruit quality characteristics. Therefore algae and two plants extracts (10% roselle and 5% garlic) could be safety recommended as a natural biostimulants application for improving most desirable measurements of mango cv. Fagri Kalan grown under the same experimental condition.

REFERENCES

1. FAOSTAT (2013) Food and Agriculture Organization of the United Nation (FAO). http://www.fao.org.
2. Abd El-Migeed AA, El-Sayed AB, Hassan HSA(2004) Growthenhancementof olivetransplants by brokencellsoffreshgreenalgae assoilapplication. Minufia J Agric. Res 29: 723-737.
3. Abd El-Moniem EA,Abd-Allah ASE(2008) Effectofgreenalgaecellsextractasfoliarspray on vegetativegrowth,yieldandberrie squalityofsuperior grapevines. Am. Euras, J Agric. and Environ. Sci 4: 427-433.
4. Spinelli F, Giovanni F, Massimo N, Mattia S,Guglielmo C(2009) Perspectives on the use of a sea weed extract to moderate the negative effects of alternate bearing in apple trees. J Hort. Sci. Biotechn 17: 131-137.
5. Ahmed FF, Mansour AEM,Montasser MAA,Merwad MA,Mostafa EAM(2013) Response of Valencia orange trees to foliar application of roselle, turmeric and seaweed extracts. J of Applied Sciences Research 9: 960-964.
6. Bruneton J(2001) Farmacogenosia. Zaragoza (Ed.)Acriba, pp.294-296.
7. Parakash L,Majeed S(2003) Multifunctional ingredients.The Nova Face of Natural C&T 118: 41-47.

8. Pons LI(2003) FotoproteccionVegeta (II).Offarm 22: 163-164.
9. Sayeeda F, Ahmad MU(2005) Comparative efficacy of some organic amendment and a nematicide (furadN-3G) against rootknot on two local varieties of groundnut. J PlantPathol 4: 54-57.
10. El-Shayeb NSA (2009) Physiological studies. Oenotherabiennis (Bio-fertilizer and plant extracts) Ph.D.Thesis, Hort. Dept, Fac. Agric, Benha University, Egypt.
11. Abbas MJ, Jerry AN, Fayadh MH(2007) Effect of plant extracts and growth regulators on endogenous hormones and carbohydrate of snak cucumber (CucumismelovarFlexuesesNaud) and cucumber (Cucumissativus L.) J Kerbala University Iraq, 5: 274-283.
12. Ahmed FF,Morsy MH(1999) A new method for measuring leaf area in different fruit species. Minia J of Agric. & Developpp.97-105.
13. Pregel F (1945) Quantitative Organic Micro Analysis (4th Ed) J A. Churchill Ltd. London pp: 53.
14. Champman HD and Pratt PE(1961) Methods of Analysis for Soil, Plant and Water. Davis Agric. Sci. Pull Office Calif. Univ.pp.220 -308.
15. Brown JD,Lilleland D (1946) Rapid determination of potassium and sodium in plant material and soil extract by flame photometer. Proc. Amer. Soc. Hort. Sci. 48: 331-346.
16. Dubois M, Gilles MA,Hamilton JK,Rebers PA,Smith F(1956) Colorimetric method for determination of sugar related substances. Anal Chem. 28:350-356.
17. Association of Official Agricultural Chemists (1995) Official Methods of Analysis A. O. A. C.(15thedn.) Published by A. O. A. C. Washington D.C. USA, pp. 490-510.
18. Snedecor W,Cochran WG(1989) Statistical Methods(8thedn.) Iowa State Univ. Press Ames. Iowa. USA.
19. Wallar A,Duncan DB(1969) Multiple ranges and multiple tests. Biometrics 11:1-24.
20. Mstat-C(1989) Users guide a microcomputer program for the design, management and analysis of agronomic research expirements. Michigan University, East Lansing, MC, USA.
21. Chawdhury MNA, Rahim MA, Khalequzzaman KM, Humauan MR,Alam MM(2007) Effect of plant extracts and time of application on incidence of anthracnose, yield and quality of mango. IntJ Sustain Crop Prod 2: 59-68.
22. Karim MA, Rahim MA(2008) Effects of different plant extracts and chemicals on yield and quality of fruits of four mango varieties. J Agrofor. Environ. 2: 113-117.
23. Abd- El- Motty- Elham Z, Shahin MFM, El-Shiekh MH, Abd El-Migeed MMM(2010) Effect of Algae extract and yeast application on growth, nutritional status, yield and fruit quality of Keittemangotrees. Agric. Biol. J N. Am. pp.421-429.

24. Ahmed FF, Kamel MK,Ibrahim HIM (2014) The synergistic effects of using plant extracts and salicylic acid on yield and fruit quality of keitte mango trees. Stem Cell 5: 30-39.

25. Adel Abed El Hamied S(2014) Improving growth and productivity of "Sukkary" mango trees grown in North Sinai using extracts of some brown marine algae, yeast and effective microorganisms. 1-Mineral content of leaves and fruit growth aspects. Middle East J of Agriculture Research 3: 318-329.

26. Hegab MY, Shaarawy AMA, El-SaidaSAG(2005) SAG. Effect of Algae extract and mono potassium phosphate on growth and fruiting of Balady orange trees. Under Press in Bull. Fac. of Agric Cairo. Univ. 56: 107-120.

27. Hafez OM, Saleh MA, El-Lethy SR(2013) Response of some seedlings olive cultivars to foliar spray of yeast and garlic extracts with or without vascular arbuscularmycorrhizal fungi.J World Applied Sciences24: 1119-1129.

7

CHAPTER

PROPAGATION PRINCIPLES IN USING INDOLE-3-BUTYRIC ACID FOR ROOTING ROSEMARY STEM CUTTINGS

MA Elhaak[1], MZ Matter[2], MA Zayed[2] and DA Gad[2]

[1]Faculty of Science, Tanta University, Egypt

[2]Faculty of Science, Minufiya University, Egypt

ABSTRACT

Indol-3-butyric acid (IBA) was applied in 0-100 ppm concentrations as a soaking medium of rosemary cuttings for one, three and six hrs or in low (0-20 ppm) concentrations continuously for cuttings rooting. Cuttings number of induced roots and root length and rosmarinic acid, phenolic and alkaloids contents, giving rosemary plant its edible and pharmaceutical importance, were studied. Results revealed that rooting ability of cuttings was improved as a number of induced roots andby time of cutting soaking in the different concentrations of IBA. Optimal root number, which was five times the control value was achieved by soaking cuttings for three hours in 60 ppm IBA. Length and weight of induced roots were significantly varied by both time of soaking and concentration of IBA. Rosemary cuttings length slightly increased (4%) by six hours of soaking in 40 ppm IBA while their leaves areas increased (20%) by one hour of cuttings soaking in 60 ppm IBA. Leaves chlorophyll a, b and carotenoids were not affected significantly by IBA concentrations, while time of soaking was effective in increasing chlorophyll a. Rosmarinic acid was increased (5.6%) by 60 ppm IBA, but soaking in 80 ppm for one and three hours decreased it (32% and 33%). Cutting mortality during transplantation into soil was improved by low concentration of IBA or by short time of Soaking in high IBA concentration. Study preferred soaking cuttings in low concentrations of IBA especially 5 ppm continuously as it was found effective in cuttings rooting and vigor that will insure success of plant growth by such cuttings schema.

Citation: Elhaak MA, Matter MZ, Zayed MA, Gad DA (2015) Propagation Principles in Using Indole-3-Butyric Acid for Rooting Rosemary Stem Cuttings. J Horticulture 2:121. doi:10.4172/2376-0354.1000121.

INTRODUCTION

Rosemary “Rosmarinus officinalis “ belongs to the family lamiaceae (labiatae) and is native to southern Europe and the Mediterranean area, and northwestern Spain. Traditionally, the Romans used rosemary to strength memory functions and scholars take rosemary during examinations to improve memory and concentration and people have regarded rosemary as the herb of remembrance because of the plant richness in volatile oils [1]. Also, rosemary oil is used in medicine for improving bad memory, reducing headache, tension, insomnia, fever and respiratory system diseases. It is used as a natural antimicrobial and antifungal against E. coli , Pseudomonas , Aspergillus, Staphylococcus and an insecticide and as a fragrant repellant [2]. Rosmarinic acid has potential in the treatment of toxic shock syndrome, whilst the flavonoid diosmin is reputedly more effective than rutin in reducing capillary fragility. Rosmarol, a compound from the leaves, has shown remarkably high antioxidant activity [3].

The whole rosemary plant is antiseptic, antispasmodic, aromatic, astringent, cardiac, carminative, cholagogue, diaphoretic, emmenagogue, nervine, stimulant, stomachic and tonic [4-6]. Rosemary is used as a cardiac stimulant, and for the cure of indigestion, flatulence, common cold, rheumatism and dandruff [7]. An infusion of the flowering stems made in a closed container to prevent the steam from escaping is effective in treating headaches, colic, colds and nervous diseases [5].

Cutting is a well-known common and relatively cheap method used in the propagation of many ornamental plant species. It overcomes the difficulties of propagation by plant seeds. Induction of adventitious roots on cuttings is governed by the complex interaction of several factors that could be classified under two major sections. The first would involve the stock mother plants’ physiological status and their environmental conditions in addition to treatments applied to the mother plants’ themselves (e.g. etiolation, girdling and spraying with chemicals) [8]. The second includes factors concerned with the post harvested cuttings which basically include both chemical and mechanical treatments imposed on cuttings [e.g. wounding, centrifugation and growth regulator application] [8]. Physical factors such as cutting length, cutting type, stem diameter, presence or absence of a leaf and stock plant from which the cuttings are taken, play an important role in rooting ability of cuttings [9].

It has been widely documented that auxins promote adventitious root development of stem cuttings through their ability to promote the initiation of lateral roots premordia and to enhance transport of carbohydrates to the cutting base needed for root growth [8,9]. The purpose of treating cuttings with auxins is to increase the percentage of rooting, root initiation, number and uniformity of rooting [10]. It also accelerates the translocation of nutrients from upper part of the cuttings to their basal ends by increasing the activity of enzymes. This increases hydrolysis of carbohydrates for providing enough energy in rooting respond of the cells [11]. As reported by Al-Barazi and Schwabe [10], occasionally IBA treatment seems to stimulate cell division in the ray cells between the primary bundles which improved root initiation and increased uniformity of rooting.

Stem cuttings of rosemary (Rosmarinus officinalis) were treated with IAA, IBA and NAA. Soaking and quick dipping are the best treatments for initiation of early rooting, root length and survival percentage under field conditions. The quick dip method was found most effective and time saving [12]. Rosemary have a good rooting ability, and the effect of the root-forming solution induces better results, both in terms of the rooting percentage and the qualitative features of the new root system [13]. Deen and Mahmoud [14] stated that auxins stimulates root formation of rosemary if applied in a dilute concentration but higher dosage (0.2%) vigorously caused inhibitory effect in rooting of rosemary cuttings. Rosemary branches were collected at the end of each season (spring, summer, autumn and winter) and the end of winter was the best season for rooting cuttings [15]. IBA is a synthetic root-promoting chemical compound that has been found most reliable in stimulating rooting of cuttings in a large number of plant species and is non-toxic to plants over a wide concentration range [8].

The present study aims at studying the effect of applying different high concentrations of IBA for different time or low concentration for long time on the characteristics of the induced roots and morphology, growth and chemical constituents of rooted cuttings.

MATERIALS AND METHODS

The present work was carried out in laboratory of Physiology, Botany department, Faculty of science Minufia University. It was carried out on rosemary (Rosmarinus officinalis L.). Cuttings of rosemary were collected from shrubs which planted in the faculty garden in clay soil.

Effect of IBA Concentration and Soaking Time

Different concentrations of IBA (Merck) (0, 20, 40, 60, 80 and 100 ppm) were prepared. IBA was firstly dissolved in few drops of 1N KOH before diluting by distilled water to the needed solution concentration. Cuttings of rosemary, fixed at about 15 cm in length by cutting under water, were sterilization by washing with HgCl2 (0.1%) and washing several times with sterilized distilled water. The lower leaves (at the lowest 3 cm) of the rosemary cuttings were removed and then two cuttings were immersed in 50 ml of each IBA concentration in beakers (50 ml) for three fixed times (one, three or six hours). The lowest three cm were totally immersed in the hormone solution. After that, cuttings were transferred into another beakers containing 50 ml sterilized distilled water after washing with distilled water. Three beakers were used as replicates for each concentration. The beakers were covered with aluminum foil to keep lower part of cuttings in continuous darkness to enhance root initiation as auxin works efficiently in darkness. The water in the beakers was kept at specific level by adding distilled water and the beakers were agitated from time to time for gently enriching the water with air. The experimental beakers were kept under laboratory conditions, air temperature 30 ± 2°C, relative humidity 80%, with 12 hours natural light and 12 hours dark. Died cuttings were recorded and excloded while others were harvested for measurements and analysis at the end of experiment.

Effect of Continuous Soaking in Low Concentrations of IBA

Low concentrations of IBA (0, 1, 2, 3, 5, 10, 15 and 20 ppm) were prepared and in 50 ml of each concentration two prepared rosemary cuttings were soaked, three beakers replicates for each concentration. The beakers were agitated from time to time and their solutions were completed to the specific level by distilled water or the same concentration. The beakers were covered with aluminum foil for darkness and kept under the previously mentioned laboratory conditions. Then died cuttings were recorded and excluded while others were harvested for measurements and analysis at the end of experiment.

When no new root was produced experiments were ended and at that time cuttings of rosemary were collected and washed with distilled water. Then the induced roots on each cutting were counted and the lengths (cm) of the roots were measured, while cuttings length (cm) at start and end of experiments was measured from the end of terminal node to the end of cutting. Fresh leaf samples from cuttings of each treatment were used for the measurements of leaf area and the photosynthetic pigments and rosmarinic acid. The roots and cuttings were weighted as fresh and dried in an electric oven at 70°C for constant weight for the determination of water content percentage. The dried materials were powdered and reserved in paper bags for further analysis.

Estimation of Photosynthetic Pigments

Fresh leaf samples (100 mg) were homogenized immediately in 5 ml cold aqueous acetone 85% (v/v) solution and centrifuged for 15 minutes at 3000 rpm. These extracts were diluted to the appropriate volume (10 ml) with the previous cold acetone solution and their color intensities were measured against a blank of the 85% acetone solution at three wavelengths, 663, 644 and 452.5 nm by using a spectrophotometer (Metertek SP-850) for the determination of chlorophyll a, chlorophyll b and carotenoids content according to the method of Metzner et al. [16].

Estimation of Rosmarinic Acid

The method of rosmarinic acid extraction and determination was reported by Lopez-Arnoldos et al. [17] and Komali and Kalidas [18]. Rosemary fresh leaf samples (200 mg) were blended with porcelain mortar in 10 ml of 50% methanol and placed in water bath at 55°C for 2 hours then were centrifuged for 10 minutes at 3500 rpm. One ml of the extract was diluted with 9 ml of the 50% methanol and the absorbance of the formed colour was measured at 333 nm using spectrophotometer (Metertek SP-850) and the rosmarinic acid concentration (mol-1cm-1) was calculated from the following equation:

$$A = \varepsilon bc$$

Where: (A) is the absorbance at 333 nm, (C) is the concentration of rosmarinic acid, (ε) is the extinction coefficient (ε=19000 L mol-1 cm-1) and (b) is the width of cuvette (b=1 cm). The content per mol was calculated as mg/g F.wt. by multiplying with rosmarinic acid molecular weight.

Estimation of Flavonoids

The aluminum chloride colorimetric method was used to estimate the flavonoids in rosemary. The rosemary cuttings dry leaves sample (10 mg) was extracted by 80% ethanol then a definite volume of solution (0.5 ml) was mixed with 1.5 ml of 95% ethanol, 0.1 ml of 10% aluminum chloride, 0.1 ml of 1M potassium acetate and 2.8 ml of distilled water. After incubation at room temperature for 30 minutes, the absorbance of the reaction mixture was measured at 415 nm with a spectrophotometer (Metertek SP-850). The amount of 10% aluminum chloride was substituted by the same amount of distilled water in blank. Quercetin was used to make the calibration curve, the reacted flavonoids of the extracts with aluminum chloride was used for the determination of flavonoids content as mg/g d.wt. in the plant leaves

Estimation of Total Alkaloids

Total alkaloids of rosemary cuttings dry leaves were. The plant leaves dry powder samples (1 g) were extracted several times with 4:1 of 70% (v/v) ethanol and glacial acetic acid. The mixture was left to stand for at least 6 hours extraction time during which it was shacked from time to time, and filtrated. The alkaloids content in the collected supernatants was precipitated by a drop wise addition of concentrated ammonia solution until no precipitation was formed. The precipitated alkaloids were filtered on a pre-weighted filter paper (Whatman 102), washed with diluted ammonia and were dried in an oven at 70°C to a constant weight. Alkaloids content was calculated as mg/g dry weight of the plant samples.

Estimation of Saponins

The saponin content in the rosemary cuttings was. A known weight (10 mg) of the plant leaves dry powder was extracted three times with 95% ethanol. The clear supernatants were combined and completed into a definite volume, then 0.5 ml of this ethanol extract was mixed with 0.5 ml of 8% vanilline in ethanol (8 g of vanilline dissolved in 100 ml ethanol). The mixture was placed in an ice bath and mixed with 5 ml of 72% sulfuric acid, then heated in a water bath adjusted at 60°C for 10 minutes followed by cooling in the ice-cold water bath. The absorbance of the formed colour was measured at 544 nm using the previously mentioned spectrophotometer. A standard curve by different concentrations of cholesterol was constructed as in the previous steps and used for the determination of the content of saponins (mg/g d.wt) in the plant leaves.

Table 1: Mortality percentage (%) of rosemary cuttings which soaked in different concentrations of IBA for different times. Mortality percentage (%) of rosemary cuttings which soaked continuously in different concentrations of IBA

Time of soaking (hour) Concentration (ppm)	One	Three	Six
0	38.8	38.8	38.8
20	33	0	33.3
40	17	0	33.3
60	17	17	33.3
80	0	0	33.3
100	0	50	0
Mean	17.5	17.5	28.6
Number of cutting	36	36	36

Estimation of Total Phenolic Compounds

Total phenolic content in rosemary cuttings was estimated quantitatively using a method. A known weight (10 mg) of the dried plant leaves was extracted by 95% ethanol three times and the clear supernatants were combined and completed to a known volume (25 ml) by the 95% ethanol. Then 1 ml from this extract was mixed with 1 ml folin reagent and 1 ml N (20% w/v), then the mixture was completed up to a known volume with distilled water. Thereafter, the absorbance of the produced colour was measured spectrophotometerically (Metertek SP-850) at 650 nm after exactly 30 minutes. A standard curve was prepared by using different concentrations of pyrogallol as the previous procedure and used for the determination of the total phenolic compounds content (mg/g d.wt.) in the plant leaves.

Statistical Analysis

The obtained results were statistically analyzed using one and two ways analysis of variance (ANOV) to determine the degree of significance for the obtained variations by the used treatments. Also, correlation coefficients were applied for investigating the significance of the relationships between the studied variables of the study plant. All of the statistical methods were according to Bishop [19], while the analysis was carried out by SPSS statistical package.

Results I- Effect of different indole-3-Butyric Acid Concentrations for Different Times

a- Root characteristics

Root number: Results of induced root number per cutting of rosemary (Table 1 and Figure 1) showed remarkable and significant variations in the number of the induced roots of cutting by time of Soaking and the concentration of the IBA.

Soaking in the different concentration of IBA for one, three and six hours increased root number progressively in comparison with the control. The maximum root number after Soaking for one, three and six hours was at 100, 60 and 40 ppm IBA concentration respectively these maxima were 17.8,27.8 and 22.5 roots and they were 223.6%,405 and 309% the control value respectively. This showed that root number was lowered but not lower than that of the control by cuttings soaking in increased concentrations of IBA also, increasing time of soaking increased root number of cutting at lower IBA concentration.

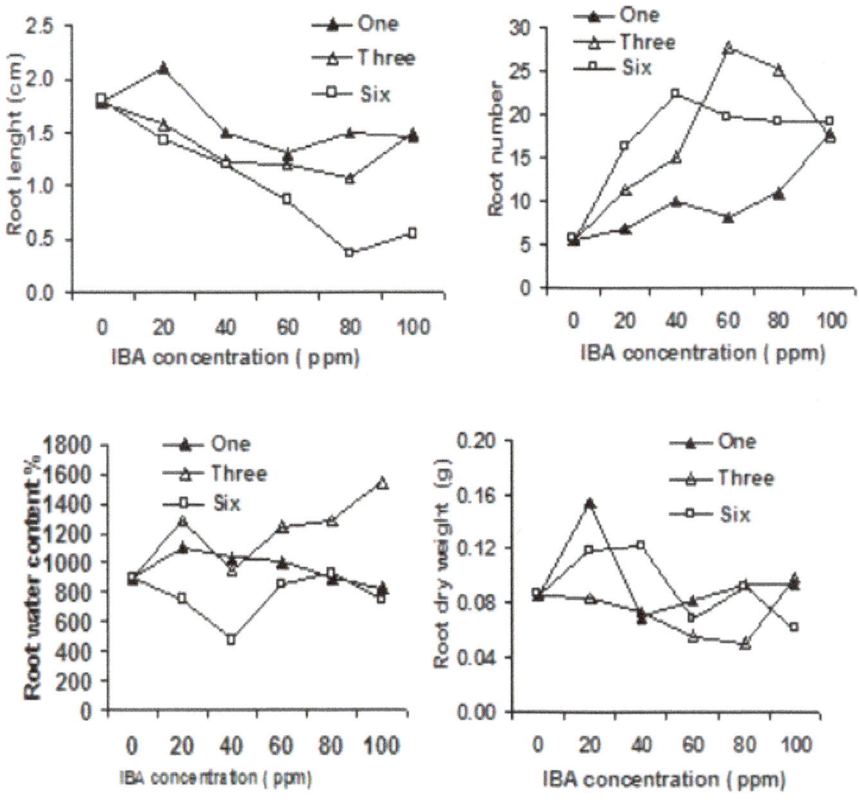

Figure 1: Variations in the induced roots number, length (cm), water content (%) and dry weight (g) on the rosemary cuttings treated with high concentrations of IBA for different times.

Root length: The lengths of the induced roots on the rosemary cuttings by concentration and time of Soaking in IBA treatments were slightly varied (Table 1 and Figure 1). Increasing time of cuttings soaking in IBA decreases root length in rosemary cuttings. Decreases were also, recorded in root length of cuttings with increasing the concentration of IBA when compared with the control, except presoaked in 20 ppm IBA for one hour which increased the cuttings root length by about 17% in comparison with the control. The shortest root lengths were 1.3, 1.1 and 0.4 cm by Soaking in IBA for one, three and six hours in 60, 80 and 80 ppm concentrations respectively. These values were lower by 33.3%, 38.8% and 77.7% than the control value for the previous Soaking treatments respectively.

Root dry weight: The exhausted metabolites in rooting were represented by produced root dry weight under the different IBA treatments (Table 1 and Figure 1). The dry weights of the induced roots on the rosemary cuttings were significantly varied by concentrations of IBA. High dry weight of roots was due to Soaking rosemary cuttings for one hour, followed by six hours in the different concentrations of IBA which progressively increased dry weights of induced roots specially with increasing the concentration of IBA, except 40 and 60 ppm for one hour and 60 and 100 ppm for six hours soaking treatments which decreased the dry weight in comparison with the control. The highest root dry weight was by Soaking for one hour in 20 ppm where the root dry weight was 1.8 that of the control. Soaking for three hours caused decreases in the roots dry weight with increasing IBA concentrations except 100 ppm where the dry weight of roots was higher by 15% than the control value.

Root water content: The percentages of water content in roots of the cuttings (Table 1 and Figure 1) were significantly varied in response to time of cuttings soaking in IBA regardless of the concentration. High percentage of roots water content was induced by soaking rosemary cuttings for three hours in the different concentrations of IBA, followed by one hour and then six hours of soaking. Soaking in most concentrations of IBA increased the percentages of roots water content when compared with the control. The maximum root water content (1544%) was recorded by soaking for three hours in 100 ppm IBA concentration and that was higher by 72% than the control value. On the other side, prolonged Soaking for six hours in all IBA concentrations, except 80 ppm concentration, decreased the percentage of water content of cuttings root compared with the control.

b- Cuttings characteristics

Cuttings length: The length of rosemary cuttings at the end of experiment was not significantly varied by Soaking time of cuttings or by the different high IBA concentrations (Table 1 and Figure 2). Slight increase was found in the length of cuttings by soaking for one, three and six hours in 80 ppm of IBA compared with the control. These increases were only by 3.3%, 2.8% and 4% of the control value.

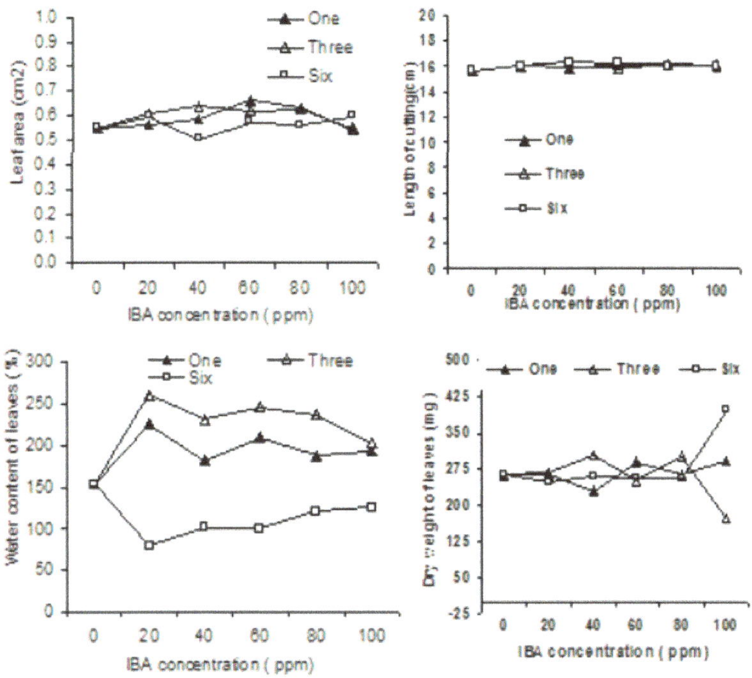

Figure 2: Variations in the cuttings length (cm) and cuttings leaves area (cm2), water content (%) and dry weight (G) of the rosemary cuttings treated with high concentrations of IBA for different times.

Cuttings leaf area: The leaf area of rosemary cuttings varied significantly by soaking for different times in the different high concentrations of IBA but was not changed significantly by concentrations of IBA (Table 1 and Figure 2). Recorded increase in the cuttings leaf area by concentrations of IBA was slight as compared with the control. The maxima of cuttings leaf area (0.66, 0.64 and 0.60 cm^2) were exhibited by soaking in 60, 40 and 20 ppm concentrations of IBA for one, three and six hours respectively. The maximum values of leaf area of rosemary cuttings were greater by more than 20%, 16% and 9% than the control value by the previous treatments of IBA respectively.

Cuttings leaf dry weight: The dry weight of rosemary leaves (Table 1 and Figure 2) was higher in the presoaked cuttings for six hours than in the presoaked for one and three hours in the different IBA concentrations. The maximum dry weights of leaves (293, 305 and 397 mg/g dry weight) were recorded by soaking cuttings in 100, 40 and 100 ppm IBA for one, three and six hours respectively which increased leaves dry weights by 11%, 16% and 51% respectively in comparison with the control value.

Cuttings leaf water content: Percentages of leaves water content significantly varied as a response to time of cutting Soaking in different concentrations of IBA (Table 1 and Figure 2). Higher water content in the leaves of rosemary cuttings was in presoaked cuttings for three hours than for one or six hours in all IBA concentrations. Six hours as a period of soaking decreased percentage of water content under all IBA concentrations compared with the control value. The highest water content (260%) was recorded in soaked cuttings in 20 ppm IBA for three hours that and was higher by about 70% the control value.

Cuttings leaf photosynthetic pigments: The content of photosynthetic pigments (chlorophyll a, b and carotenoids) in the leaves of rosemary cuttings varied significantly ($p<0.05$) by time of cuttings soaking in IBA but not significantly with IBA concentrations (Table 1 and Figure 3).

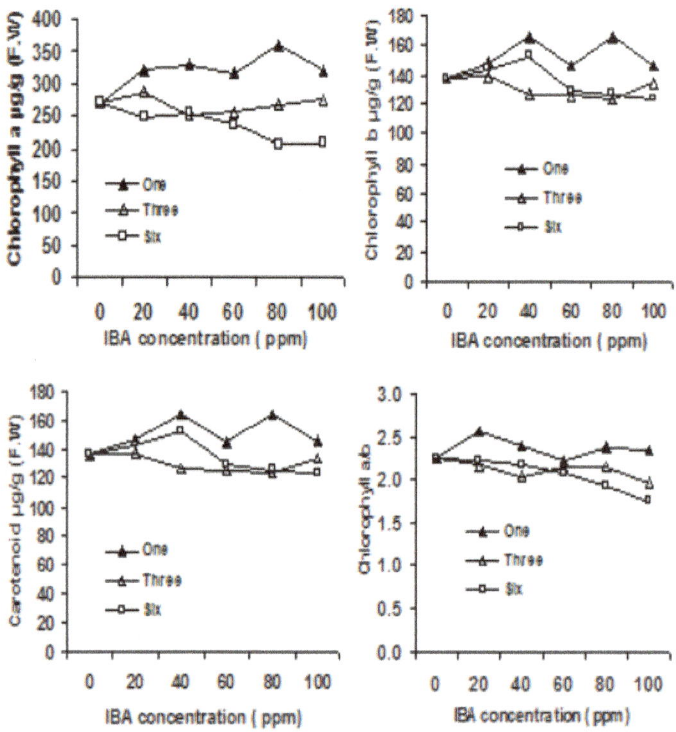

Figure 3: Variations in cuttings leaves chlorophyll a, b and carotenoids content (mg/g f.wt) and the ratio of chlorophyll la/b of the rosemary cuttings treated with high concentrations of IBA for different times.

The content of chlorophyll a in the leaves of rosemary cuttings decreased in case of elongation of the soaking time of in the different concentrations of IBA. The soaked cuttings for one hour in most concentrations of IBA showed high chlorophyll a content compared to the control. Prolonged soaking time to three and six hours, on the other hand, led to a fluctuation of chlorophyll a contents around a decreasing trend by increasing the concentrations of IBA. The maximum chlorophyll a content (360 µg/g fresh weight) was due to soaking cuttings for one hour in 80 ppm IBA concentration and was higher than the control by 33%.

On the contrary the content of chlorophyll b in rosemary cuttings leaves which was greatly lower than chlorophyll a content decreased due to increasing time of cuttings soaking in the different concentrations of IBA (Table 1 and Figure 3). Chlorophyll b content increased in response to the different concentrations of IBA except for six hours soaking in most of the used concentrations. Soaking in 80 ppm IBA for one hour led to the highest chlorophyll b content (151 μg/g fresh weight) which was higher by 26% than the control value.

The carotenoids content in the leaves of rosemary cuttings was greater by short time of soaking (one hour) than by the long ones (three and six hours) in IBA concentrations (Table 1 and Figure 3). Also, progressive increases in carotenoids content was exhibited by increasing the concentration of IBA up to 80 ppm especially for one hour soaking time. All concentrations of IBA led to higher carotenoids than compared to the control, especially when cuttings were presoaked for one hour, while increasing time of Soaking for three and six hours mostly decreased carotenoids content than that of the control. The highest carotenoids content (165 μg/g fresh weight) was by 80 ppm soaking for one hour and was higher by 20% the control value.

Chlorophyll a/b: Chlorophyll a/b of rosemary leaves was slightly higher by soaking for one hour than for three and six hours (Table 1 and Figure 3). Chlorophyll a/b ratios ranged between 1.74 and 2.56. The highest ratio was due to one hour soaking in 20 ppm that was higher by 13% than the control value. Soaking for three and six hours decreased the ratio especially with increasing the concentrations of IBA. The decrease in chlorophyll a/b ratios was due to more decrease in chlorophyll a than in chlorophyll b by the time of Soaking.

Cuttings leaf rosmarinic acid: The rosmarinic acid contents in the leaves of rosemary cuttings varied significantly ($P<0.01$) by IBA concentrations, time of soaking and their interaction (Table 1 and Figure 4). Soaking for one, three and six hours decreased rosmarinic acid content of cuttings specially with increasing the concentration of IBA when compared with the control value, except the slightly higher content than that of the control by Soaking for six hours in 20 and 60 ppm IBA concentrations. The recorded increases in rosmarinic acid content (by 20 and 60 pmm) were only greater by about 1.2% and 5.6% than the control value. The marked decreases (32% and 33% of the control value) were recorded by soaking for one and three hours respectively in 80 ppm.

Cuttings leaf Flavonoids: The results of flavonoids content in the leaves of rosemary cuttings (Table 1 and Figure 4) varied significantly ($P<0.01$) by concentrations of IBA, time of Soaking and their interactions. The flavonoids content in the leaves of rosemary cuttings was decreased by increasing time of cuttings soaking in IBA. But, increasing the concentration of IBA led to slight decreases in flavonoids content of cuttings when compared with the control. Decreases in flavonoids content ranged between a minimum decrease by 0.32% to a maximum decrease by 42.3% in comparison with the control value and after Soaking for one hour in 100 ppm and six hours in 100 ppm of IBA.

Cuttings leaf Alkaloids: Alkaloids content in leaves of rosemary cuttings (Table 1 Figure 4) showed significant decrease ($P<0.01$) by both IBA concentration and time of Soaking. Soaking for one hour increased alkaloid conte nt of cuttings by increasing the concentration of IBA when compared with the control. But higher than 40 ppm,

IBA attenuated alkaloids content gradually. The maximum alkaloids content was 21.4 mg/g dry weight at 40 ppm and that was higher by 27.8% than the control. Also soaking for three hours increased alkaloids with increasing the concentration of IBA in comparison with the control value and its maximum (higher by 29%) was recorded by 100 ppm. Alkaloids content was increased after six hours of soaking in 80 ppm of IBA concentration (17.85 mg/g), while it decreased with the other used concentrations compared to control value.

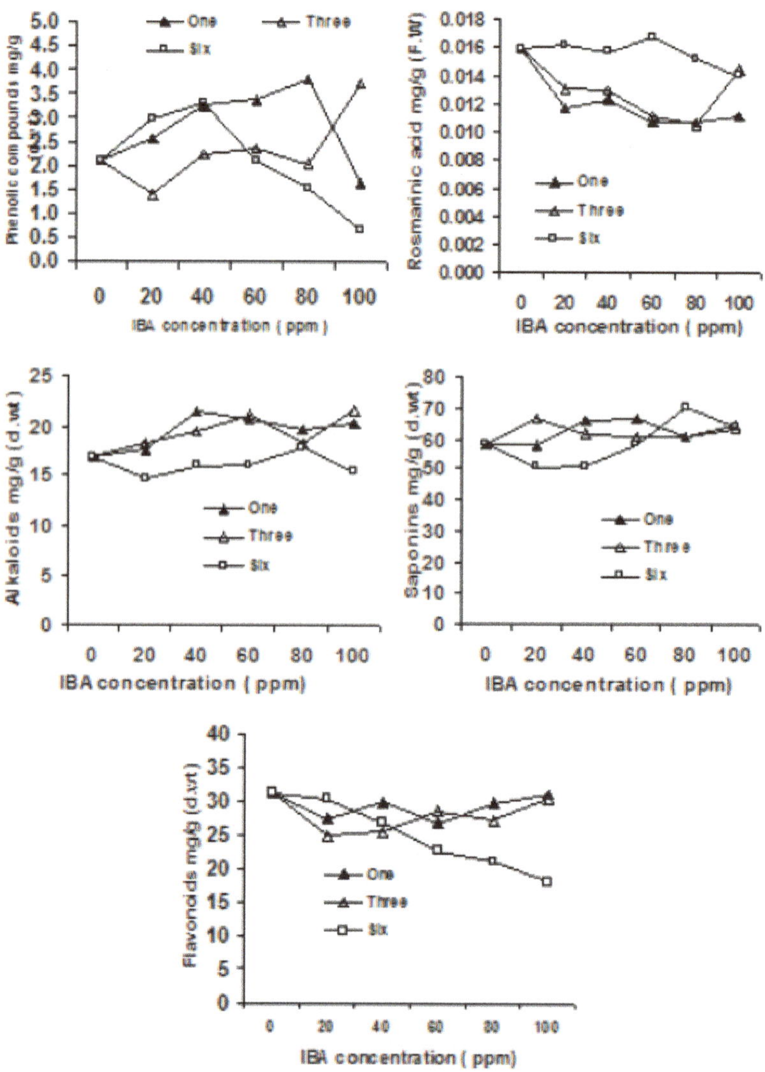

Figure 4: Variations in cuttings leaves phenolic compounds, rosemarinic acid, alkaloids, saponin and flavonoids contents (mg/g d.wt) of the rosemary cuttings treated with high concentrations of IBA for different times.

Cuttings leaf Saponins: Saponin content in leaves of rosemary cuttings did not vary significantly by Soaking in different concentrations of IBA and the different times of Soaking (one, three and six hours) or their interaction (Table 1 and Figure 4). But there were general decreases in mean saponins content by soaking time from one to six hours. Soaking for one hour increased saponins content of cuttings with increasing the concentration of IBA up to 60 ppm then the content was decreased gradually but still higher than the control value. The maximum saponins content was 66.5 mg/g at 60 ppm and was higher by 14.4% than the control value. Three hours Soaking increased also saponins content with increasing the concentrations of IBA in comparison with the control and the maximum content was higher by 15% and was caused by 20 ppm concentration of IBA. Soaking for six hours in IBA concentrations more than 60 ppm increased saponins content with a maximum at 80 ppm and was higher by 20.7%, then the content was decreased but still higher than the control value.

Cuttings leaf Phenolic compounds: The phenolic compounds content in the leaves of rosemary cuttings significantly varied with the Soaking in the different concentrations of IBA ($P<0.01$), the time of soaking ($P<0.05$) and the interaction of the two applied treatments ($P<0.01$) (Table 1 and Figure 4). A general decrease of phenolic compound content in the leaves of rosemary cuttings was exhibited by elongating time of soaking in IBA. Phenolic compounds contents ranged between a maximum of 3.83 and a minimum of 0.64 mg/g dry weight by Soaking in 80 and 100 ppm of IBA for one and six hours respectively. Soaking for one hour, in increasing the concentration of IBA increased phenolic compound content compared with the control value up to the 80 ppm then decreased the content to lower than the control value by 100 ppm IBA concentration. Soaking for three hours led to fluctuation of the content of phenolic compound around an increasing trend with increasing the concentration of IBA. The maximum content (3.72 mg/g dry weight) was by 100 ppm and was higher by 75.4% compared to the control. Soaking for six hours was accompanied with increase in the phenolic compounds content with increasing IBA concentrations up to 40 ppm, which increased the content by 57%, and followed with a sharp decrease with higher IBA concentrations.

II- Effect of continuous soaking in low concentration of IBA

a- Root characteristics

The experimental results of continuously soaked rosemary cuttings in the low concentration of IBA (Figure 5) significant increase in root number by increasing concentrations of IBA up to 15 ppm where the maximum increase was by 1172% in comparison with control then root number was decreased but still higher than the control value. Root length also showed a general increase with the increase in IBA concentration up to 3 ppm where the maximum increase was recorded and it was by 170% compared to the control. After that the increase was slightly attenuated but root length still higher than that of control.

Fresh weight of the produced roots on the cuttings was not significantly varied. But at low concentrations (1, 2 and 3 ppm IBA) there were decreases, and at the higher concentrations of IBA there were varied increases and its maximum was recorded by the highest IBA concentration (20 ppm) and was by 309% compared to the control. Dry weight of the cuttings root was not significantly varied with the concentration of IBA. Remarkably the root dry weights were lower than that of the control. Water content of root exhibited significant increase with the increase in IBA concentration with a maximum increase at 3 ppm and was by 712% compared to the control.

a- Cuttings characteristics: Length of rosemary cuttings which continuously soaked in low concentration of IBA was not changed significantly by the different concentrations of IBA (Figure 5). The observed increase in comparison with the control cutting length did not exceed 3.2%. Area of leaves of rosemary cuttings also was not significantly by IBA in comparison with the control. There was an increase at 1 ppm and was by 50%, in comparison with the control. Fresh and dry weights of rosemary cuttings leaves (Figure 6) significantly increased with all IBA concentrations and 10 ppm caused the maximum increase (by 149% and 152% compared to control). The variation in water content of rosemary leaves was significant by all IBA concentration. IBA concentration 1 and 3 ppm increased leaf water content, the maximum at 3 ppm was higher by 17% the control value while other concentrations slightly decreased it.

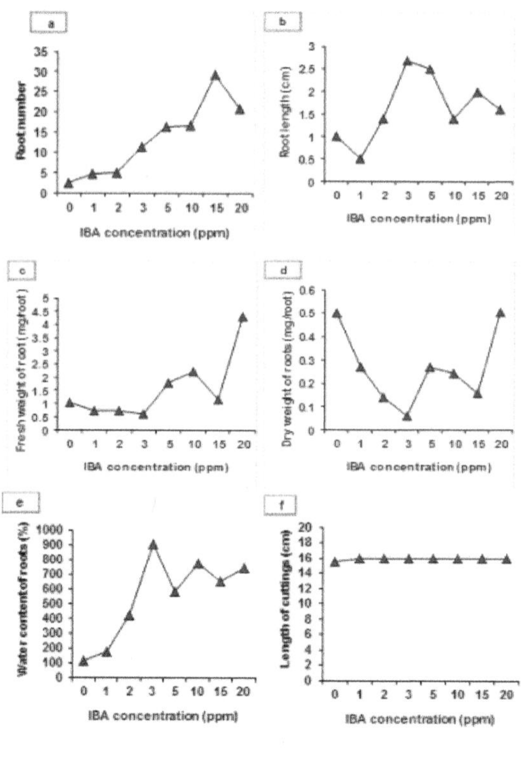

Figure 5: Effect of continuous soaking in different concentrations of IBA on (a) Root number, (b) Root length (cm), (c) Fresh weight of roots (mg/root), (d) Dry weight of roots (mg/root), (e) Water content of roots (%) and (f) Length of cuttings (cm).

Important cutting metabolites: Data of chemical analysis of leaves of rosemary cuttings which soaked continuously in low concentrations of IBA were represented in (Figure 6-8). Chlorophyll a content of rosemary leaves was not significantly varied as a response to IBA concentrations. There were slight increases over the control value by all IBA concentration except 5 ppm and the maximum increase was recorded by 20 ppm and it was by 19.4%. Chlorophyll b content was not also significantly varied but it was decreased than the control value with increasing IBA concentration with a minimum content was at 5 ppm that was lower by 26.8% than control value. Carotenoids in rosemary leaves exhibited slight increase by low concentrations of IBA (1 and 2 ppm) while by high concentrations it decreased. The minimum content was at 5 ppm and was lower by 16.8% than the control. Chlorophyll a/b ratio was higher under in IBA concentrations than under control and acquired its highest at 10 ppm. Also, chlorophyll (a+b)/c ratio slightly increased than the control ratio with the increase in IBA concentration, the highest ratio was at 3 ppm concentration. This ratio decreased compared to that of the control by 15 and 20 ppm concentrations.

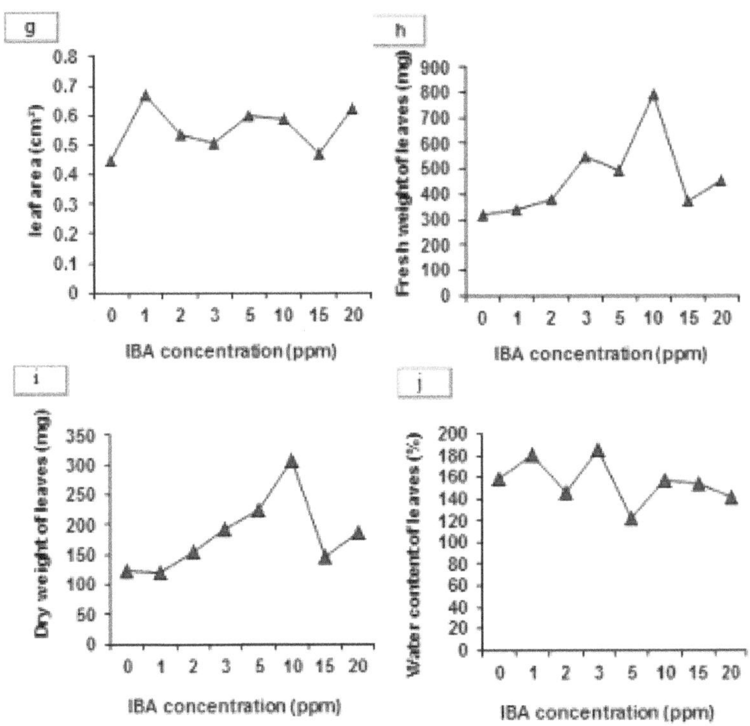

Figure 6: Effect of continuous soaking in different concentrations of IBA on (g) Leaf area (cm2), (h) Fresh weight of leaves (mg), (i) Dry weight of leaves (mg) and (j) Water content of leaves.

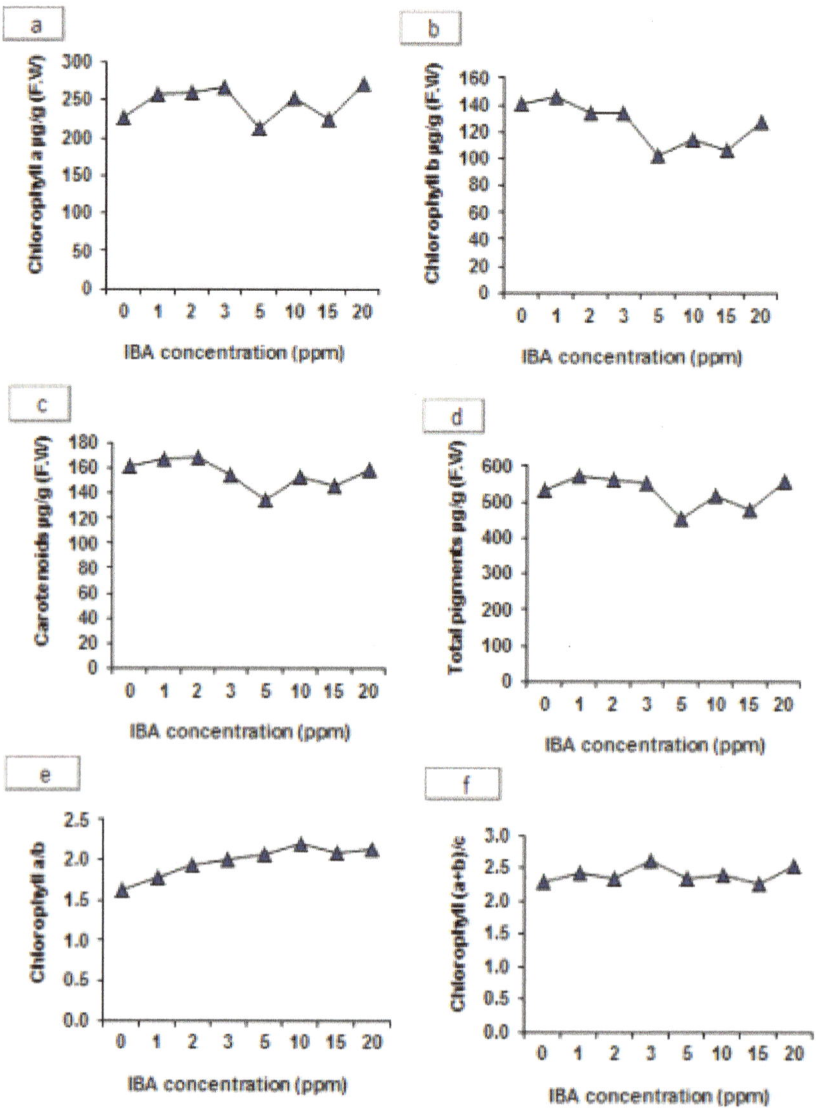

Figure 7: Effect of continuous soaking in different concentrations of IBA on (a) Chlorophyll a µg/g fresh weight, (b) Chlorophyll b µg/g fresh weight, (c), Carotenoids µg/g fresh weight, (d) Total pigment µg/g fresh weight, (e) Chlorophyll a/b, (f) Chlorophyll (a+b)/c.

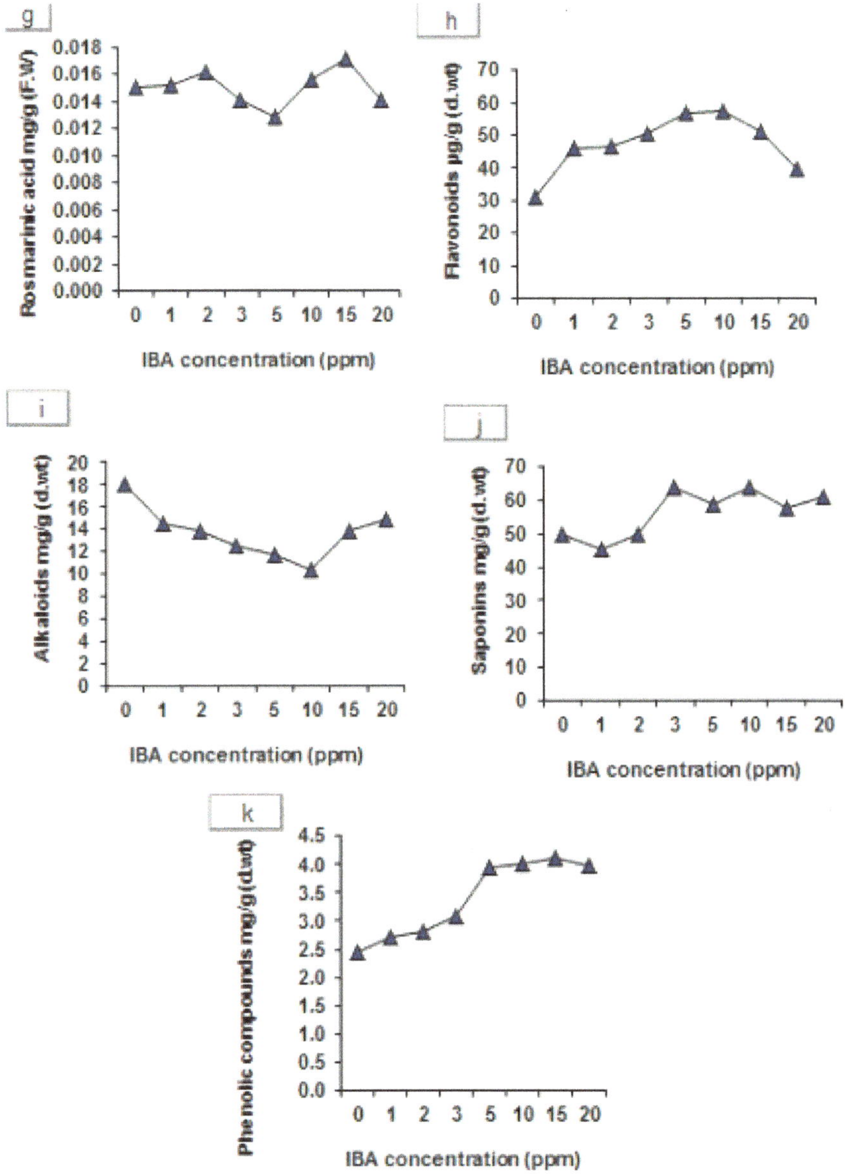

Figure 8: Effect of continuous soaking in different concentrations of IBA on (g) Rosmarinic acid mg/g fresh weight, (h) Flavonoids mg/g dry weight, (i) Alkaloids mg/g dry weight, (j) Saponins mg/g dry weight and (k) Phenolic compounds mg/g dry weight.

The rosmarinic acid content in rosemary cuttings leaves generally, did not significantly increase with the increase in concentrations of IBA. The maximum increase was at 15 ppm and was higher by 14% than the control value. The rosmarinic acid minimum content was at 5 ppm IBA concentration and was lower by 14.6% than the control. On the opposite flavonoids content in rosemary leaves significantly varied by IBA concentrations. Increased in content of flavonoids was recorded with the increase in IBA concentration up to 10 ppm where the maximum content was recorded and it was higher by 83.2% in comparison with the control. Continuous soaking of rosemary cuttings in IBA caused a marked (but not significant) decrease in alkaloids content especially with increasing IBA concentration up to 10 ppm. The minimum alkaloids content was at 10 ppm and was lower by 42.7% of the control value.

Saponins content in rosemary leaves varied significantly by IBA concentrations. At low concentrations of IBA(1 and 2 ppm) saponins content decreased than the control value but at high concentrations it increased. The maximum content of saponins was at 10 ppm IBA concentration and was higher by 27.6% than the control value. Phenolic compounds varied significantly and showed a paralleled increase with the increase in IBA concentration up to 15 ppm at which the increase was by 68.3% as compared with control.

Relationships between the Measured Parameters in Rosemary

Root number significantly and positively correlated with water content of roots ($r<0.05$), and cut length ($r<0.01$) (Table 2). Root length significantly and positively correlated with chlorophyll a ($r<0.01$) and negatively with rosmarinic acid ($r<0.05$). Dry weight of roots significantly and positively correlated with root length ($r<0.01$). Water content of roots significantly and positively correlated ($r<0.01$) with only chlorophyll b. Dry weight of leaves significantly and negatively correlated with carotenoids ($r<0.05$). Water content of leaves significantly and positively correlated with root length, chlorophyll a, chlorophyll b and with leaf area and carotenoids ($r<0.05$). Water content of leaves correlated at $r<0.01$ with rosmarinic acid. Rosmarinic acid acquired significant negative correlations with chlorophyll a and chlorophyll b

Table 2: Relationships between the measured parameters in rosemary

Concentration (ppm)	0	1	2	3	5	10	15	20	Number of cutting
IBA continuous	66.6	66.6	50	50	0	16.6	50	50	48

The correlation matrix between morphological and chemical characters of rosemary cuttings under the effect of soaking continuously in different concentrations of IBA indicated that root number significantly and positively correlated with root length, phenolic compounds ($r<0.01$) and water content of roots ($r<0.05$). Root length correlated significantly and positively ($r<0.05$) with leaf area and negatively ($r<0.05$) with chlorophyll a. Fresh weight of roots correlated significantly and positively with dry weight of roots ($r<0.01$) and leaf area and Saponins ($r<0.05$). Dry weight of roots significantly and positively correlated with leaf area (r<0.05). Water content of roots significantly and positively correlated with saponins (r<0.01), fresh weight of leaves and phenolic compounds (r<0.05) and negatively with flavonoids and alkaloids (r<0.05). Fresh weight of leaves significantly and positively correlated (r<0.01) with dry weight of leaves and negatively correlated at r<0.05 with flavonoids and alkaloids. Dry weight of leaves significantly and positively correlated (r<0.05) with phenolic compounds and negatively (r<0.05) with alkaloids. Chlorophyll a significantly and positively correlated (r<0.01) with chlorophyll b, carotenoids and total pigments. Chlorophyll b significantly and positively correlated (r<0.01) with total pigments. Carotenoids correlated significantly and positively at r<0.01 with total pigments.

Mortality of Cuttings During Rooting and their Survival after Transplantation

Mortality percentage of rosemary cuttings was increased with increasing time of Soaking and decreased with increasing IBA concentrations (Table 3). The mortality of cuttings reached zero by Soaking in the concentrations of IBA (20, 40, 60, 80 and 100 ppm). The greatest mortality was under the control (38.8%). Mortality of cuttings which soaked continuously in IBA decreased with increasing concentration until it reached zero by 5 ppm IBA after it mortality increased again with the increase in IBA concentration, which suggesting this concentration for storage of hormone and could be used for reserving the plant cuttings during manipulation as transportation before cultivation.

The transplanted rooted cuttings into peat moss soil survive and continue to growg. All the IBA treatments led to 100% survival of the rooted cutting in the peat moss soil with out any other additions. The continuous observation of the transplanted cuttings for three months indicated more vigor and better growth for IBA treated cuttings in comparison with the control ones.

Table 3: Correlation coefficient (r) values of the relationships between rosemary morphological and chemical characters under the effect of presoaking in different concentrations of IBA at different times. Root number (RN), Fresh weight of root (FWR), Dry weight of root (DWR), Water content of root (WCR), Fresh weight of leaves (FWL), Dry weight of leaves (DWL), Water content of leaves (WCL), Cut length (Cut L), Root length (Root L), Leaf area (Leaf A), Rosmarinic acid (RA), Phenolic compounds (Ph), Saponins (Sapo), Flavonoids (Flavo), Alkaloids (Alka), Chlorophyll a (Chl a), Chlorophyll b (Chl b), Carotenoids (Carot), Total pigments (TP), (*) Correlation is significant at r<0.01

	RN	FWR	DWR	WCR	FWL	DWL	WCL	CutL	Root L	Leaf A	RA	Ph	Sapo	Flavo	Alka	Chl a	Chl b	Carot	TP
RN	1																		
FWR	-0.011	1																	
DWR	-0.08	-671(**)	1																
WCR	-228(*)	-558(**)	-0.129	1															
FWL	0.048	0.021	-0.145	0.137	1														
DWL	0.107	-0.106	-0.212	0.053	-697(**)	1													
WCL	-0.091	0.146	0.028	0.139	-509(**)	-0.212	1												
Cut L	-471(**)	0.045	0.124	-0.085	0.044	0.18	-0.125	1											
Root L	0.099	-300(**)	-425(**)	0.074	0.175	-0.144	.397(**)	0.048	1										
Leaf A	-0.011	-0.088	-0.161	-0.001	0.073	-0.081	.247(*)	-0.07	0.063	1									
RA	-0.103	0.024	0.171	-0.059	-.531(**)	-0.152	-.574(**)	-0.092	-.329(*)	-0.234	1								
Ph	-0.155	0.134	0.25	-0.132	-.369(**)	-.276(*)	-0.176	-0.043	0.096	0.024	-0.014	1							
Sapo	0.054	-0.234	-0.232	-0.152	-0.067	-0.142	0.125	-0.071	-0.166	-327(*)	-0.164	-0.083	1						
Flavo	0.173	0.021	0.26	-0.156	-.562(**)	0.017	-.776(**)	0.208	-.402(**)	-0.258	.757(**)	0.072	-0.073	1					
Alka	0.078	-0.012	-0.246	0.132	0.191	-0.077	.351(*)	-0.049	0.137	0.158	-.488(**)	0.192	0.209	-.594(**)	1				
Chl a	-0.239	0.088	-0.096	0.23	0.258	-0.159	.553(**)	-0.216	.366(**)	0.256	-.527(**)	0.242	-0.02	-.676(**)	-472(**)	1			
Chl b	-0.072	0.084	-0.251	-358(**)	0.245	-0.131	.517(**)	-0.21	0.194	0.246	-.459(**)	0.117	0.091	-.597(**)	-475(**)	-843(**)	1		
Carot	-0.16	0.112	-0.011	0.193	-0.072	-.304(*)	.286(*)	-0.232	0.189	0.252	-0.183	-375(**)	0.032	-.292(*)	.306(*)	-804(**)	-752(**)	1	
TP	-0.196	0.097	-0.122	0.267	0.204	-0.191	.522(**)	-0.23	.310(*)	0.268	-.472(**)	0.252	0.017	-.619(**)	-467(**)	-980(**)	-912(**)	-878(**)	1

DISCUSSION

The study was designed to investigate the effect of indole-3-butyric acid (IBA) hormone that might initiate the rooting of rosemary cuttings needed for rosemary plant propagation. Cuttings of rosemary were subjected to high concentrations of indole-3-butyric acid (IBA) for different times or to low concentrations continuously to initiate roots on the cuttings. At the time of no new roots were produced the cutting were harvested and the morphological and metabolical characteristics of the produced roots and of cuttings were studied and the interrelationships between these characters were analyzed.

The number, length, fresh and dry weights, and water content of the induced roots were affected by IBA treatments as also found by Nada et al. [20]. Soaking in different concentrations of IBA for one, three and six hours had various impacts on the morphological characteristics of the rosemary cuttings which are in agreement with the findings of Panwar et al. [21] and Schoellhorn [22]. The number of the induced roots per cutting of rosemary was significantly increased by time of cutting Soaking and concentration of the IBA as also found by Talia et al. [13], Scalon et al. [23] and Nada et al. [20]. The root number was five times the control value by Soaking cuttings in 60 ppm IBA for three hours but greater root number was induced on cutting by increasing time of Soaking in lower IBA concentration as was found by Shah et al. [24] and Silva and Pedras [15]. This increase in root number and length by IBA improve the success and survival of the cuttings after transplanting in the soil. Also, Hartmann et al. [8] appreciated using IBA in stimulating rooting as it is non toxic to plants over a wide concentration range and it improves root initiation and uniformity of rooting. The length of the induced roots on rosemary cuttings was sensitive to both IBA concentration and time of soaking in. however, reduction in root length was exhibited by IBA concentrations and elongation of soaking as was also found by Chauhan et al. [12] while, low concentrations increased the cuttings root length and for example 20 ppm increased root length by about 17%. The marked reduction in root lengths was due to six hours of soaking in the highest IBA concentration (100 ppm), data was in contrast with that of Scalon et al. [23].

The dry weight of induced roots on rosemary cuttings acquired high values the used concentrations of IBA at three hours of soaking but those values accompanied with a decrease in root number. Similar results were also found by Silva and Pedras [15]. This may indicate accumulation in metabolites in the produced root instead of initiation of new ones. Soaking for three hours at 100 ppm of IBA increased dry weight of roots by 15%. Similar effects have been reported by Hosni et al. [25] and Singh et al. [26,27] on Bougainvillea buttiana and by Rowezak [28] on Ficus retusa and F. benjamina. The IBA changes of root fresh and dry weights reduced significantly the water content of the induced root. The response was remarkable by time of cuttings soaking in IBA but slightly by IBA concentrations. This may shortened the time of cuttings tolerance during cultivation as roots will dry earlier.

The growth of rosemary cuttings was slightly during soaking in IBA where the cuttings length was increased by 3.3%, 2.8% and 4.0% compared with the control value by one, three and six hours of Soaking in 80, 80 and 40 ppm IBA concentrations respectively. Similar slight increase was in leaf area of rosemary cuttings by time of soaking in the used concentrations of IBA that reached 20, 16.3 and 9% over the control plant leaf area by one, three and six hours of cuttings soaking in 60, 40 and 20 ppm IBA concentrations respectively. This may indicate that short time of Soaking in high concentration of IBA induces cuttings growth and increases their leaf area which is important for cutting propagation. This also may indicate that metabolism process of cuttings was not only directed by the hormone to root initiation but also to the growth of cuttings. This was also confirmed by the fresh weight of rosemary cuttings leaves which was increased by soaking for one and three hours (33.8% maximum increase), But on the contrary, soaked cuttings for six hours in most IBA concentrations inhibited their growth as was found by Azimi and Bisgrove [29] in Rosa. This inhibition in cuttings growth due to the soaking for six hours caused accumulation of metabolites in the cuttings that increased their dry weight as compare with one and three hours soaking. The effect was magnified by soaking in high concentration of IBA (100 ppm) which led to the high leaves dry weight confirming observed reduction in dry material utilization in root production and or growth in cuttings. These responses were in agreement with the findings of Singh et al. [26] and Sharma et al. [30]. Devlin and Witham [31] attributed the promotion effect of IBA for vegetative growth to the enhancement of rooting and root growth on the treated cuttings that enhances uptake of water and nutrients from the growing medium. The water content percentages in rosemary leaves significantly responded to changes in time of cutting soaking in the different concentrations of IBA [31]. High leaves water content percentages were recorded in the presoaked cuttings for one and three hours in all IBA concentrations and prolonging time of Soaking to six hours remarkably decreased the leaves water content.

The photosynthetic pigments content of rosemary leaves was not affected significantly by IBA concentrations, but time of Soaking in these IBA concentrations was significantly effective as soaking for one hour caused a marked increase in chlorophyll a content which is in agreement with the results of Abu-Grab and Ebrahim [32] and Abdel- Wahed et al. [33]. According to Midan et al. [34], Ludwig- Muller et al. [35] and Ludwig-Muller [36], IBA increases chlorophyll accumulation. Soaking for longer time 3 and 6 hrs) decreased chlorophyll a with progressive values by the increase of IBA concentrations. The minimum content was due to Soaking for six hours in this IBA concentration, results agreed with those of Wiesmann and Lavee [37]. Chlorophyll b metabolism enhancement by one hour of cutting Soaking was also attenuated by the increase in soaking time for three and six hours which decreased chlorophyll b especially with the highest concentration of IBA. Similar effects have been reported by Wiesman and Lavee [37], in the cuttings of olive. Leaves carotenoids markedly accumulated by Soaking for one hour with increasing the concentration of IBA but prolonged soaking to six hours, mostly decreased carotenoids than the control value. The total pigments in the leaves of rosemary cuttings followed the trend of variations in chlorophyll a however, one hour Soaking in all IBA concentrations led to the highest total pigments while 3 and 6 hours soaking decreased with few exceptions the total pigment content as also obtained before [37]. Chlorophyll (a+b)/c ratio increased by soaking for one and three hours in all concentrations of IBA but prolonged time of Soaking for six hours on the other hand decreased the ratio. This was due to more inhibition in metabolism of chlorophyll a and b than in that of carotenoids.

Rosmarinic acid accumulated in rosemary leaves by about 1.2% and 5.6% due to 20 and 60 ppm IBA concentrations as also found by Munne-Bosch et al. [38] and Ain-Lhout et al. [39] but it was sharply inhibited by higher concentrations soaking for one and three hours (32% and 33%). Frankel et al. [40] and Del Bano et al. [41] due the decrease in rosmarinic acid in the cuttings to the oxidation of caffeic acid by phenolases and peroxidases which is responsible for the decomposition and disappearance of rosmarinic acid, in one hand, and to the limitation of important nutrients such as nitrate and ammonium ions in the medium on the other hand.

However, flavonoid content in the leaves of rosemary cuttings decreased in response to both IBA concentration and time of soaking. Decreases in flavonoids content ranged between a minimum decrease by 0.32% to a maximum decrease by 42.3% due to soaking for one and six hours in 100 ppm of IBA. It is therefore quite possible that the decrease in flavonoids may be due to their transformation into IAA in adventitious root formation as reported by Stenlid [42] and Mosella and Macheix [43].

Both IBA concentration and time of Soaking affect significantly alkaloids of leaves of rosemary cuttings with an exception of one hour soaking that increased alkaloids especially when concentration of IBA increased. The maximum alkaloids content was 21.4 mg/g dry weight. Three hours soaking increased alkaloids with increasing the concentration of. Alkaloids content was reduced after six hours of soaking with most of used IBA concentrations.

Saponin content in leaves of rosemary cuttings exhibited a general inecrease by soaking time and with most of IBA concentrations. The maximum saponin content was 66.5 mg/g dry weight at 60 ppm and was higher by 14.4% than the control value. These results agreed with those of Kim et al. [44]. The later author found in Panax ginseng indole-3-butyric acid at 25 μM with methyl jasmonate at 100 μM stimulated ginsenoside saponin accumulation compared with 100 μM methyl jasmonate alone.

Phenolic compounds was found in the plant leaves by small countities compared with the other secondary metabolites and they acquired a general decrease in the leaves of rosemary cuttings by elongating time of soaking in IBA. Phenolic compounds contents ranged between a maximum of 3.83 and a minimum of 0.64 mg/g dry weight. Increase in phenolic compounds by IBA was reported by many authors such as Rowezak [28], Habba [45], Qaddoury and Amssa [46] and Amin et al. [47].

The continuous soaking of rosemary cuttings in low concentrations of IBA increased the produced roots and this increase was a function of concentration of IBA. Maximum rooting was by 15 ppm which increased roots about 10 times compared to the control. These results resembled the results of Shah et al. [24] and Silva and Pedras [15]. The length of produced roots on cuttings increased in response to the increase in IBA concentration. The longest roots were due to 3 ppm concentration of IBA where root length was increased by 170% compared to control. This result is in agreement with those of Scalon et al. [23].

Fresh weight of growing roots on the cuttings increased with high concentrations of IBA. Similar effect had been reported by Hosni et al. [25] and Singh et al. [26,27] on Bougainvillea buttiana and Rowezak [28] on Ficus retusa and Ficus benjamina . Dry weight of the roots decreased with increasing the concentration of IBA in comparison with control. This marked decrease in root dry weight could be a result of utilization of metabolites in the growth of roots length. Water content of root increased with the increase in IBA concentration.

Length of rosemary cuttings and area of their leaves did not show marked effect with continuous application of IBA. This indicated 1- A negligible affect of long application of low concentrations of IBA on the morphology of the cutting 2- Over all metabolites were directed to rooting. These results agreed with those of Hussein [48]. Fresh and dry weight of rosemary cuttings leaves significantly increased with IBA concentrations. The maximum increase was at 10 ppm. The increase in the vegetative growth of rosemary due to IBA treatments was also found by Singh et al. [21] on Gardenia lucida . Sharma et al. [30] reported that IBA improved plants vegetative growth.

The photosynthetic pigments chlorophyll a, b and carotenoids of rosemary leaves was affected by continuous soaking of cuttings in IBA. Chlorophyll a metabolism was accelerated by the increase in IBA concentration with a maximum at 20 ppm that was higher by 19.4% compared to control. These results are in agreement with those obtained by Abu-Grab and Ebrahim [32] and Abdel-Wahed et al. [33]. The metabolism of chlorophyll b and carotenoids on the opposite was mostly inhibited and the resulted decreases reached to 26.8% and 16.8% at 5 ppm concentration of IBA. Similar effects have been reported by Wiesman and Lavee [37] and Abu-Grab and Ebrahim [32]. Total pigments were increased with IBA concentrations as a result of the marked increase in chlorophyll a over the decrease of the other pigments. This was also confirmed by the increase in chlorophyll a/b and chlorophyll (a+b)/c ratios. The highest increase in the previous ratios was in chlorophyll a/b especially at 10 ppm where it was by 37.5%.

The prolonged application of the used low concentrations of IBA had no effect or slightly increased rosmarinic acid of rosemary leaves. The maximum increase (14% compared to control) was at 15 ppm. Which agree with results of Reda et al. [49]. Continuous soaking of rosemary cuttings in IBA inhibited the metabolism of alkaloids which decreased by 42.7% at 10 ppm IBA concentration. Creus and Barcklo [50] found slight decrease in the alkaloids content by IBA on Nicotiana rustica

The marked increase in response to prolonged IBA application was in the phenolic compounds of rosemary leaves. The 15 ppm concentration of IBA led to 68.3% increase compared to control. Qaddoury and Amssa [46] reported that phenolic compounds increased immediately after IBA treatment. Saponin in rosemary leaves accumulated only at high concentrations of IBA and Kim et al. [44] found stimulation in saponin accumulation by similar treatments.

The number and character of the induced roots by IBA at specific time correlation coefficients showed significant correlations between cut length, water content, chlorophyll a, total pigments and rosmarinic acid. Also, the content of rosmarinic acid in the rosemary leaves was found to be sensitive to the water content and photosynthetic pigments of the cuttings. Mortality of cuttings which soaked continuously in IBA decreased with increasing concentration until it reached zero by 5 ppm IBA after it mortality increased again with the increase in IBA concentration, which suggesting this concentration for reserving the plant cuttings during manipulation during transportation.

REFERENCES

1. Bown D (1995) Encyclopaedia of Herbs and their Uses. New York, DK Publishing, Inc., pp. 12-49.
2. Duke JA (2001) Hand Book of Medical Herbs. Boca Raton, FL: CRC press, p. 677.
3. Duke JA, Ayensu ES (1985) Medicinal Plants of China. Reference Publications, Inc., ISBN: 0-917256-20-4.
4. Lust J (1983) The Herb Book. Bantam books, pp. 50-79, ISBN: 0-553-23827-2.
5. Grieve M (1984) A Modern Herbal. Penguin, ISBN: 0-14-046-440-9.
6. Polunin O, Huxley A (1987) Flowers of the Mediterranean. Hogarth Press, pp. 12-44, ISBN: 0-7012-0784-1.
7. Chakravarti S,Raghuvanshi SS (2005) Rosemary (Rosemarinusofficinalis): a useful medicinal herb. Vaniki-Sandesh, 29: 26-27.
8. Hartmann HT, Kester DE, Davies FT (1990) Plant Propagation: Principles and Practices. 5th ed., pp. 1-100. Prentice Hall International Editions, Englewood Cliffs, New Jersey, USA.
9. Leakey RRB (1983) Stock plant factors affecting root initiation in cuttings of Triplochitonscleroxylon K. Schum., an indigenous hardwood of West Africa. Journal of Horticultural Science, 58: 227-290.
10. AL-Barazi A,Schwabe WW (1982) Rooting softwood cuttings of adult Pistaciavera. Journal of Horticultural Science, 57: 247-252.
11. Arya S, Tomar R, Tokoyt OP (1994) Effect of plant age and auxin treatment on rooting response in stem cuttings of Prosopis cineraria. Journal of Arid Environments, 27: 99-103.
12. Chauhan VK, Jagmohan S,Srivastava LJ (1992) Initiation of rooting in stem cuttings of rosemary through hormonal treatments. Indian Journal of Forestry, 15: 131-135.
13. TaliaMAC, Viola F,Forleo LR (2004) Vegetative propagation of two species of the Mediterranean maquis (Rosmarinusofficinalis L., Viburnum tinus L.) for applications in naturalistic engineering. Italus-Hortus, 11:89-92.
14. Deen SE, Mahmoud M (1996) Comperative study between saponin and natural auxin on root growth of Rosemary (Rosmarinusofficinalis) cutting. ActaHorticulturae, 426, 635-642.
15. Silva C,Pedras JF (1999) Early rooting in rosemary (Rosmarinusofficinalis L.) cuttings under the influence of chemical treatments and collecting time. Acta-Horticultrae, 502, 213-217.
16. Metzner H, Rau H,Senger H (1965) UltersuchungenZursynchronisierbarkeiteinzelner pigment mangel - Mutanten von chlorella. Planta 65: 186-194.

17. Lopez-Arnaldos T, Lopez-serrano M, Barcelo AR, Zapata JM (1995) Spectrophotometric determination of rosmarinic acid in plant cell cultures by complexation with Fe2+ ion. Fresenius. J Anal Chem 351-314.
18. Komali AS, Kalidas S (1998) Comparison of the growth pattern and rosmarinic acid production in rosemary shoots and genetically transformed callus cultures. Food Biotechnology, 12: 27-41.
19. Bishop ON (1983) Statistics in Biology. Longman, Penguin, London, pp. 56-63.
20. Nada Paradikovic, SvjetlanaZeljkovic , Monika Tkalec, VinkovicT, Irma Dervic, et al.(2013) Influence of rooting powder on propagation of sage (Salvia officinalis l.) and rosemary (Rosmarinusofficinalis l.) with green cuttings. Poljoprivreda 19: 10-15.
21. Panwar RD, Gupta AK, Yamdagni R, Saini RS (1999) Effect of growth regulators on the rooting of cuttings of Bougainvillea cv. Thimma. Haryana AgricUniv J Res 29: 11-17.
22. Schoellhorn R (2001) Effect of growth regulators on rooting and stem elongation in propagation of five flowering tropical perennials. ActaHort 559, 43-48.
23. Scalon SPQ, Ramos MBM,Vieira, MdoC (2003)Auxins and boron on length of the biggest root and number of rooted cuttings of Guaco (MikaniaglomerataSprengel), Rosemary (Rosmarinusofficinalis L.) and Carqueja (Baccharistrimera Less A.P.D.C) in two periods of planting. Revista-Brasileira-de-Plantas-Medicinais, 5: 71-76.
24. Shah SC, Gupta LK,Bhujwan HS (1996) Effect of auxins on rooting of stem cuttings of rosemary (Rosemarinusofficinalis Linn.). Recent Horticulture, 3: 126-128.
25. Hosni AM, El-Gendy SA, Shedeed MR, Ebrahim AK (2000) Improvement of rooting in Bougainvillea buttiana 'Mrs Butt' by wounding and/or IBA application(s) to cutting basal ends. AnnAgricSci 45: 659-678.
26. Singh HP, Kohli RK,Batish DR (2001) Allelopathy inagroecosystems: an overview. J Crop Prod 4, 1–41.
27. Singh P, Tewari N,Katiyar PK (2002)Pretransplant seedling treatment with growth regulators and their effect on the growth and bulb production of onion (Alliumcepa L.) progressive. Agric 2: 181-182.
28. Rowezak MMA (2001) Response of some Ornamental plants to Treatment with Growth Substances. M.Sc. Thesis, Fac. Agric., Cairo Univ., Egypt.
29. Azimi M,Bisgrove RJ (1975) Rooting of hardwood cuttings. ExpHortic 27: 22-27.
30. Sharma AK, Trivedi ON,Shukla PK (2002) Effect of IAA and IBA on Acalypha cuttings. J OrnHort 5: 60-62.
31. Devlin RM, Witham FH (1986) Plant Physiology, (4thedn.) CBS Publishers and Distributors, Delhi, India, p. 54, 57, 292 and 381.
32. Abu-Grab OS,Ebrahim MKH (2000) Physiological response of field-grown onion to some growth regulators. Egypt J Hort 27: 117-130.

33. Abdel-Wahed MSA, Amin AA, Rashed ME (2006) Physiological effect of some bio regulators on vegetative growth, yield and chemical constituents of yellow maize plants. World J of Agric 2: 149-155.

34. Midan AA, El-Bakry AM,Malush NM (1982) Growth, chemical constituents and yield of onion in relation to growth regulators application. Res Bull FacAgricZagazigUniv, Egypt (508).

35. Ludwig-Muller J, Sass S, Sutter EG, Wodner M, Epstein E (1993) Indole-3-butyric acid in Arabidopsis thaliana 1. Identification and quantification. Plant growth Regul 13: 179-187.

36. Ludwig-Muller J (2000) Indole-3-butyric acid in plant growth and development. Plant Growth Regul 32: 219-230.

37. Wiesmann Z, Lavee S (1995) Enhancement of IBA stimulatory effects on rooting of olive cultihvar stem cuttings. Hort Science, 62, 189-198.

38. Munne-Bosch S, Alegre L, Schwarz K (2000) The formation of phenolic diterpenes in Rosemarinusofficinalis L. under Mediterranean Climate. Eur Food ResTechnol 210, 263-267.

39. Ain-Lhout F, Diaz Barradas MC, Zunzunegui M, Rodriguez H, Garcia Novo F, et al. (2004) Seasonal differences in photochemical efficiency and chlorophyll and carotenoid contents in six Mediterranean shrub species under field condition. Photosynthetic, 42: 399-402.

40. Frankel EN, Huang SW, Prior E, Aeschbach R (1996) Evaluation of antioxidant activity of Rosemary extracts, carnosol and carnosic acid in bulk vegetable oils and fish oil and their emulsions. Journal of The Science of Food and Agriculture, 72: 201-208.

41. Del Bano AJ, Lorente J, Castillo J, Benavente-Garcia O, Del Rio JA, et al. (2003) Phenolic diterpenes, flavones, and rosmarinic acid distribution during the development of leaves, flowers, stems, and roots of Rosmarinusofficinalis. Antioxidant activity. Journal of Agricultural and Food Chemistry, 51, 4247-4253.

42. Stenlid G (1962) The effect of some synthetic anthocyanidins on growth and ion absorption of roots. Physiol Plant, 15, 598-605.

43. Mosella LC, Macheix JJ (1979) Le microbouturage in vitro du Pecher (PrunuspersicaBatsch): Influence de certains composes phenoliques. C R AcadSci Paris, 289, 567-570.

44. Kim Y, Yeung EC, Hahn E, Paek K (2007) Combined effects of phytohormone, Indole-3-butyric acid, and methyl jasmonate on root growth and ginsenoside production in adventitious root cultures of Panax ginseng C.A. Meyer. Biotechnology Letters, 29: 1789-1792

45. Habba EE (2003) Physiological studies of some growth regulators on the growth, yield and chemical constituents of onion plants (Allium cepa L.) JAgricSci Mansoura Univ 28: 1645-1653.

46. Qaddoury MA, Amssa M (2004) Effect of exogenous indole butyric acid on root formation and peroxidase and indole-3-acetic acid oxidase activities and phenolic contents in date Palm offshoots. Bot Bull Acad Sin 45, 127–131.
47. Amin AA, Rashad ME, Gharib FAE (2006) Physiological response of maize plants (Zea mays L.) to foiliar application of morphactin CF125 and indole-3-butyric acid. J of BiolSci 6: 547-554.
48. Hussein MMM (2008). Effect of planting dates and indole-butyric acid on rooting of Beaumontiagrandiflora, Wallich. Cuttings and consequent plant growth. Arab Univ. J Agric Sci. Ain Shams Univ. Cairo, 11: 765-787.
49. Reda F, Abel-Rahim EA, El-Baroty GSA, Ayad HS (2005) Response of essential oils, phenolic components and polyphenol oxidase activity of thyme (Thymus vulgaris, L.) to some bioregulators and vitamins. International Journal of Agriculture and Biology 7: 735-739.
50. Creus JA, Barcklo J (1988) Effect of indole-3-butric acid, kinetin, gibberellic acid on alkaloid content in Nicotianarustica stem cuttings.BiologiaPlantarum, 30:104-110.

8
CHAPTER

INCREASING FRESH EDAMAME BEAN SUPPLY THROUGH SEASON EXTENSION TECHNIQUES

Nolen S[1], Zhang B[2] and Kering MK[1]

[1]Agricultural Research Station, Virginia State University, Petersburg, Virginia, USA
[2]Virginia Polytechnic Institute and State University, Blacksburg, USA

ABSTRACT

Soybean [Glycine max (L.) Merr.], a legume native to East Asia contains up to 40% protein and 20% oil. Edamame, a popular vegetable soybean in East Asia, especially China and Japan, harvested at reproductive stages six and seven (R6 or R7) is gaining popularity in the US. Increased awareness of its nutritional quality through promotional campaigns and changing population demographics in the US have led to recent raise in US market demand for edamame. To meet the increasing market demand, frozen edamame from China and Taiwan has been imported. However, the quality of such imported product quickly deteriorates under frozen condition. The objective of this study was to determine whether off-season production systems and staggered planting of different maturity groups (MG) edamame cultivars during the planting season can extend the harvesting window for fresh beans. Four released edamame cultivars of different maturity groups (MG) were used: Gardensoy31 (MG III), Gardensoy41 (MG IV), Mooncake (MG V) and Randolph (MG VI). Total pod yield, marketable pod yield and seed quality traits including protein, oil and sucrose content of each cultivar were determined. The results indicated that early- and mid-spring planting of all MG varieties in the high tunnels allows for pod harvest starting in early July. Planting early in plasticcovered field followed by conventional planting in late-spring allowed harvest in mid-summer through early fall. While total and marketable pod yield differed among cultivars and production systems, seeds had comparable oil content (158 g kg-1) and protein and sucrose content range of 370-422 g kg-1 and 33-73 g kg-1 on dry matter, respectively. Use of season extension production techniques and soybean of appropriate MG increases harvest window from two weeks to several months.

Citation: Nolen S, Zhang B, Kering MK (2016) Increasing Fresh Edamame Bean Supply through Season Extension Techniques. J Horticulture 3:170. doi:10.4172/2376-0354.1000170.

INTRODUCTION

Japanese vegetable soybean (edamame) is harvested at reproductive (R) growth stage six to seven (R6 to R7) when the seeds are still green [1]. At this developmental stage, the seed contains high protein, monounsaturated fatty acids, minerals elements, as well as vitamins like B1 and B2 [2,3]. Edamame, a historically Japanese delicacy has been rapidly incorporated in to American diets due to changing ethnic diversity among US population demographics and increased people's awareness of edamame nutritional qualities. Edamame is currently found in grocery stores and farmers' markets with its demand in the U.S. being estimated at 14,877 tons annually [4,5]. Edamame production is a promising opportunity to meeting the specialty crop market demands. While over 70% of the edamame sold in supermarkets across the U.S is imported from China and Taiwan, it is mainly marketed frozen and this frozen storage conditions lowers its quality drastically. Commercial frozen edamame may contain harmful bacteria such as Escherichia coli and Listeria monocytogenes [6]. In addition, consumers raise the concern of edamame source of safety and start to seek locally grown fresh edamame [7].

Lack of long-term, local supply has become a barrier for successfully marketing of edamame in the U.S because it has a very narrow field harvest window of only a few days. There is very limited information available on extending edamame harvest window though season extension production techniques in Virginia. However, growth studies under plastic film house culture and plastic film tunnel on two Japanese edamame cultivars has been done elsewhere [8]. By planting from February 15 to September 25, they were able to harvest crop from May 25 to December 3. In the US, an intensive research was conducted to extend edamame harvest window in 2003 and 2004 at University of Kentucky [9]. Four Gardensoy series edamame cultivars from MG I to IV were planted in greenhouses and transplanted into the field either covered by plastic film or with no plastic protection on April 1st, 15th, and 29th in Kentucky [9]. The crop was harvested for fresh pods from June 17th (GardenSoy 11, MG I) to August 10th (GardenSoy 41, MG IV). In the study, GardenSoy 11 had lower marketable yield than other cultivars at all transplanting dates, and indication showed that early production by transplanted MG I cultivars may not be successful in Kentucky. Because no research has been done in Mid-Atlantic region to extend fresh edamame harvest window, growers are looking for superior edamame cultivars with high stability and adaptability. There is a possibility for increased on-farm income from edamame relatedbusinesses through off-season supply. To be able to supply fresh edamame for an extended period of time, there is a need for season extension system development.

MATERIALS AND METHODS

The experiment was carried out for two growing season (2013-2014) at Randolph Research and Demonstration Farm, Virginia State University. The study was laid out as a randomized complete block design with three replications. Four edamame cultivars of different maturity groups (MG) where used: GardenSoy31 (MG III), GardenSoy41 (MG IV), Moon Cake (MG V) and Randolph (MG VI. Edamame planting in early- and mid-spring was done in a high tunnel (a polyethylene-covered semi-circular structure). Planting date varied because of unforeseen weather conditions, however, it was categorized into early-spring (Before April 21th), mid-spring (April 21th-May 21th) and late spring (May 21th-June 21th), and late summer (Aug 21th-Sept 21th). During early and mid-spring planting in the high tunnel, the four edamame cultivars; Gardensoy31, Gardensoy41 and Moon cake Randolph were grown in seed beds covered with a black plastic. During mid-spring planting in the field, edamame was grown in both plastic- covered and non-covered seed beds. Another planting in the field with or without plastic-cover was carried out in late spring. Also in late August-early September, crop was established in the high tunnel to protected crop from cold temperatures later in the fall.

During harvest, all plants were harvested and pods removed. Because cultivars of different MG were planted on the same time, harvesting was sequential. Within a production system, the first to mature were those belonging to the low MG category. Total weight of harvested pods was determined and then sorted into pods with 1, 2, 3 or 4 seeds. The weight of pods in the individual categories was then obtained. Total weight of marketable pods was obtained as weight of all pods with two or more seeds pod-1. A representative sample of the marketable pods was obtained, weighed and shelled. Number of seeds was obtained, and its fresh weight was determined. About 100 gram of the shelled seeds were obtained and stored frozen to await freeze drying. The freeze dried material was ground later and analyzed to determine oil, protein, and sucrose content. Protein content was determined using the combustion method (AOAC - Official Method 990.03) with a Vario MAX CN (Elementar Americas, Inc., Mt. Laurel, NJ, USA) as described by Association of Official Analytical Chemists [10]. Protein factor of 6.25 is used for calculation. Oil was extracted from dried and grounded samples using petroleum ether in an ANKOM XT15 Extractor, Method 2, 01-30-09 (ANKOM Technology, Macedon NJ, USA) as described by AOCS [11] Official Procedure Am 5-04. Sugars were extracted from ground sample (1 g) and analyzed by HPLC following the methods optimized by Johansen et al. [12]. Sugars in the extracts were identified by comparing their retention times with standard sugars. For quantification, trehalose was used as internal standard and the sugar concentration was expressed as g/100 g dry basis. Statistical analyses were carried out using SAS procedures.

RESULTS AND DISCUSSION

Plant Growth and Harvest Window Extension

Plants grown in high tunnel on plastic covered seed beds in both early- and mid-spring showed good growth. For mid-spring planting, while crop planted in the field on plastic-covered beds showed faster growth than those without plastic cover (Figure 1) and both were slower than that in the high tunnel. Similar use of high tunnels and other season extension techniques have been used in other horticultural crops [13,14] and its beneficial effects is a result of microclimate modification allowing crops to grow outside of their normal time schedule.

(a) (b)

Figure 1: Growth in field plants for mid-spring planting with-(a) and without (b) plastic covered seed bed.

For crops grown in early spring, pod harvest was achieved as early as the beginning of July. In early July, Gardensoy31, a MG III vegetable soybean cultivar was harvested and Gardensoy41, MG IV was harvested about two weeks later. At the time of harvest of Gardensoy41 from early-spring planting, Gardensoy31 in the high tunnel from midspring planting was ready for harvest. Using the production systems (High tunnel, plastic covered seed- beds in the field, and conventional soybean production) and planting edamame cultivars of different maturity groups simultaneously extended the harvest window from a few days common for conventionally produced crop to several months (Table 1). Within a cultivar, harvesting occurred two weeks to a month between first and second crop in the high tunnel. For the same cultivar in the field, a crop on plastic-covered beds was harvested earlier than those without plastic, because it grew faster and matured earlier. Because of differences in maturity time attributed to planting dates, cultivar's MG, and use of multiple production systems, pod harvest occurred from early July and continued through end of September/ early-October (Table 1). While differences in MG was responsible for harvest time differences among cultivars planted at the same time under a given production system, presence or absence of plastic cover was responsible for differences within a cultivar. In general, Gardensoy31 (MG III) and Gardensoy41 (MG IV) were harvested earlier than Mooncake (MG V) and Randolph (MG VI) when planted at the same time under similar production system. However, extremely cold temperature and animal pest destroyed edamame grown in late summer for two successive years. While this crop was not included in the result, there was a potential to extend harvest into early November.

Pod yield and Yield-Related Traits

For a specific planting period and for a specific production system (high tunnel, field with or without plastic covered seed-bed), total harvested pods differed with edamame cultivars. For early spring planting in the high tunnel, total pod yield varied from 4300 kg ha-1 in Gardensoy31 to over 10,000 kg ha-1 for Randolph (Table 2). Slightly higher yields were obtained during the second planting in the high tunnel in mid- spring. The increase may be attributed to increased soil and air temperatures that may have allowed for better growth conditions since this was planted 2-3 weeks later in the spring. Yield in a crop established in August and September in the high tunnel are not included due to loss of crops from destruction by Marmota monax (ground hog), an animal pest.

Table 1: Planting time and harvest period for the dLنerent edamame varieties produced under given production systems. Key ___________________Approximate period within which Gadensoy31 and Gardensoy41 where harvested in that order. --------------------- Approximate period within which Mooncake and Randolph where harvested in that order

Production System	Planting Period	July			Aug			Sept			Oct		
		Early	Mid	Late	Early	Mid	Late	Early	Mid	Late	Early	Mid	Late
High tunnel	Early-spring												
	Mid-spring												
Field/Plastic	Mid-spring												
Field/no plastic	Mid-spring												
Conventional	Late- spring												

Table 2: Comparison of total and marketable pod yield for edamame cultivars grown under the high tunnel at dLنerent planting dates

Planting period	Cultivar	Harvested pods (kg ha^{-1})	
		Total yield	Marketable yield
Early-spring	Gardensoy31	4373b	2487b
	Gardensoy41	7551ab	3915ab
	Mooncake	9838ab	5020ab
	Randolph	12957a	7506a
Mid-spring	Gardensoy31	6117ab	4120ab
	Gardensoy41	11340ab	6348ab
	Mooncake	10505ab	6053ab
	Randolph	9432ab	6545ab

In mid-spring, though not significant, marketable yield from crop established in high tunnels or in the field with plastic covered seed-bed were relatively higher than that from field without plastic cover (Table 3). Overall, Randolph produced relatively high total pod yield than other cultivars within a given production system. In open field for mid-spring planting, Randolph yield range from 14000 to 23000 kg ha-1 fresh pods (Table 3). In late spring under conventional planting, Randolph produced the largest (P=0.05) total pod yield of 15850 kg ha-1. The other three varieties produced similar total pod yields averaged at 9900 kg ha-1 (Figure 2). The marketable yield for late spring planting ranged from 63 to 77 % of total pod yield depending on variety (Figure

2). The yields for same varieties in this study were lower than those obtained by others in Mississippi [5].While marketable yield for mid-spring planting range from a low of 4120 kg ha-1 in Gardensoy31 to 6545 kg ha-1 in Randolph under the high tunnel, it ranged from 2092 kg ha-1 in Gardensoy31 to 10,814 kg ha-1 in Randolph in field grown crop (Table 3 and Figure 2). These values for marketable yield is comparable to those of other vegetable soybean obtained other studies [15,16]. The difference in yield between cultivars within a production system could be attributed to differences in cultivar yield potentials. In field grown crop, plastic covered-bed increased yield in all cultivars. Under all the systems, marketable pods (≥ 2 seeds pod-1) were between 40-75% of the total yield. The number of pods with at least two beans was proportionally higher than the other class categories and could explain the high proportion of marketable pod yield. For all varieties, there was a strong correlation between total and marketable yield (r>0.75) (data not shown). There was also strong correlation (r>0.75) between total pod yield or marketable pod yield and number of pods with one, two, or three seeds.

This would be expected since total yield include all pod while marketable consist of those with ≥ two seeds pod-1 and makes sense that as number of these pods in harvest increase, total or marketable yield also increases. Similar correlation has been shown before in other vegetable soybean varieties [17].

The oil content of fresh seed harvested from different production systems and varieties were similar (P-=0.05) and averaged 158 g kg-1 (Table 4). The sucrose content was affected by the interaction of the production system and variety (P<0.05).

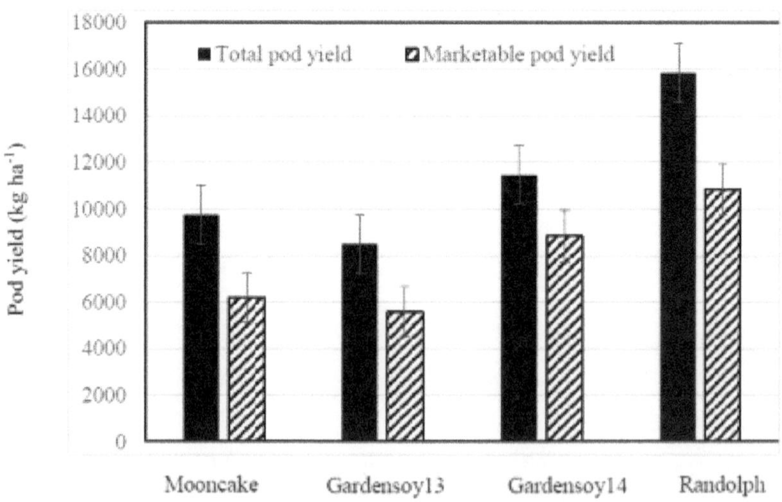

Figure 2: Total and marketable pod yield for crop under conventional planting in late spring. Values are means ± SE.

Table 3: Comparison of total and marketable pod yield in edamame cultivars planted in mid-spring using dLﻭerent production systems

Field/Plastic	Gardensoy31	9732bc	7323bc
	Gardensoy41	8788bc	6110bcd
	Mooncake	10906bc	6947bc
	Randolph	23883a	17522a
Field/no plastic	Gardensoy31	4884c	2092cd
	Gardensoy41	9163bc	5334bcd
	Mooncake	3822bc	822d
	Randolph	14310ab	10287ab

In general, sucrose content was higher for plants seeded later in the field averaging 57 g kg-1 across varieties. Across varieties, sucrose content of seed from plants in the high tunnel averaged 43 g kg-1. The protein and oil content of seed in this study were comparable to those reported elsewhere [16,18,19]. While sucrose content were comparable to those reported before [20,21], they are higher than those of vegetable varieties reported in other studies [22]. Seed protein content was affected by interaction of the production system and variety ($P<0.05$). While there was observed a general increase in seed protein content with time from mid-spring to late-spring in Gardensoy31 and Gardensoy41, there was a reduction in content for Randolph (Table 4). Across production systems, Randolph seed protein content averaged at 396 g kg-1 was larger ($P=0.05$) compared to other varieties.

Table 4: Protein and sucrose content of seeds of edamame cultivars planted at dLﻭerent times and in dLﻭerent production systems. For each seed quality attribute, values in a row followed by same lower case letter are not dLﻭerent (P=0.05), For each seed quality attribute, values in a column followed by same upper case letter are not dLﻭerent (P=0.05)

Production system	**Cultivar**	**Harvested pods (kg ha^{-1})**	
		Total yield	Marketable yield
High tunnel	Gardensoy31	6117c	4120cd
	Gardensoy41	11340bc	6348bcd
	Mooncake	10505bc	6053bcd
	Randolph	9432bc	6545bcd

Production system	Time of Planting	Variety				
		Gardensoy31	Gardensoy41	Mooncake	Randolph	Mean
		(g kg^{-1})				
		Protein				
High tunnel	Early-spring	370bB	373bC	377bB	394aAB	378C
High tunnel	Mid-spring	382bB	390bAB	422aA	408aA	401A
Field/Plastic	Mid-spring	375bB	370bC	377bB	401aAB	381C
Field/no plastic	Mid-spring	383aAB	399aA	386aB	390aB	389B
Field/no plastic	Late-spring	397aA	397aAB	374bB	387abB	389B
Mean		381b	386b	387b	396a	
		Sucrose				
High tunnel	Early-spring	63aA	46bB	33cC	44bB	46B
High tunnel	Mid-spring	37abC	35bC	40abC	45aB	39C
Field/Plastic	Mid-spring	50aB	49aB	39bC	50aAB	47B
Field/no plastic	Mid-spring	59bAB	48cB	73aA	43cB	56A
Field/no plastic	Late-spring	58aAB	59aA	59aB	55aA	58A
Mean		53a	48b	49b	48b	

CONCLUSIONS

There is great potential to increase edamame supply in Virginia using season extension techniques. Use of high tunnel and plastic covered seed-beds when ambient conditions are unfavorable for root growth creates localized soil temperatures that allow for improved plant growth. Planted early in the high tunnel, where greenhouse effects increase air temperatures allow early plant growth and early maturity. This crop will be harvested off-season during early summer and producers may access markets when vegetable edamame supply is low and a potential for higher prices exist. A longer period of harvesting time and supply of fresh beans can be achieved with selection of appropriate mix of different MG edamame varieties and production systems. However, economic analysis should be done to determine the viability and profitability of the enterprise that incorporates multiple production techniques. Collaborative work needs to be done with food packaging entities and consumer markets to determine the edamame supply and market demand dynamics so that timely production can be done.

ACKNOWLEDGEMENT

We would like to thank the Sustainable Agriculture Research and Extension (SARE) for funding this project.

REFERENCES

1. Fehr WR, Caviness CE, Burmood DT, Pennington JS (1971) Stage of development descriptions for soybeans, Glycine max (L.) Merrill. Crop Sci 11:929-931.
2. Miles CA, Chen C (2000) Washington State University Edamame trials. WashStUnivCoopExt, Vancouver, WA.
3. Kelley KM, Sanchez ES (2005) Accessing and understanding consumer awareness of and potential demand for Edamame. Hort Sci 40:1347-1353.

4. Johnson D (2000) Edamame: Westerners develop a taste for Japanese soybean. EngTechnolSustainable World 7:11-12.
5. Zhang L, Kyei-Boahen S (2007) Growth and yield of vegetable soybean (Edamame) in Mississippi. HortTechnology 17:26-31.
6. Pao S, Ettinger M, Khalid M, Mebrahtu T, Mullins C (2008) Microbiological quality of frozen "Edamame" (vegetable soybean). JFood Safety 28:300-313.
7. Montri DN, Kelley KM, Sanchez ES (2006) Consumer interest in fresh, in shell Edamame and acceptance of Edamame-based patties. HortSci 41:1616-1622.
8. Park HK, Kim KH, Oh YJ, Kim SC (2001) Selection of vegetable soybean and development of year-round production system in Korea.
9. Bec S, Pfeiffer T, Slone D (2005) Production system for extending the harvest time frame of fresh-market Edamame in Kentucky.
10. AOAC (2005) Official Method 990.03. Protein (Crude) in Animal Feed, Combustion Method, in Official Methods of Analysis of AOAC International. (18thedn),AOAC International, Arlington, VA, pp: 30-31.
11. AOCS (2005) Official Methods and Recommended Practices of the American Oil Chemists Society, Approved procedure Am 5-04 Rapid determination of Oil/ Fat utilizing high temperature solvent extraction. Urbana, IL. American Chemical Society.
12. Johansen HN, Glitso V, Knudsen KEB (1996) Influence of Extraction Solvent and Temperature on the Quantitative Determination of Oligosaccharides from Plant Materials by High- Performance Liquid Chromatography. JAgricFood Chem 44: 1470-1474.
13. Waterer D (2003) Yield and economics of high tunnel for production of warm-season vegetable crops. HortTechnology 13:339-343.
14. Carey EE, Jett L, Lamon WJ Jnr, Nennich T, Orzolek MD, et al. (1991) Horticultural Crop Production in High Tunnels in the United States: A Snapshot. HortTechnology 19:37-43.
15. Duppong LM, Hatterman-Valenti H (2005) Yield and quality of vegetable soybean cultivars for production in North Dakota. HortTechnology 15:896-900.
16. Carson LC, Freeman JH, Zhou K, Welbaum G, Reiter M (2011) Cultivar evaluation and lipid and protein contents of Virginia-grown edamame. HortTechnology 21:131-135.
17. Li Y, Du M, Zhang QY, Wang G, Hashemi M, et al. (2013) Correlation and path coefficient analysis for yield and components of vegetable soybean in northeast China. Legume Research 36:284-288.
18. Rao MSS, Bhagsari AS, Mohamed AI (2002) Fresh green seed yield and seed nutritional traits of vegetable soybean genotypes. Crop Sci 42: 1950-1958.
19. Li Y, Du M, Zhang QY, Wang G, Hashemi M, Liu X (2012) Greater differences exist in seed protein, oil, total soluble sugar and sucrose content of vegetable soybean genotypes [Glycine max (L.) Merrill] in Northeast China. AJCS 6:1681-1686.

20. LiY, Du M, Zhang Q, Wang G, Hashemi M, et al. (2014) Planting date influences fresh pod yield and seed chemical compositions of vegetable soybean. HortScience 49:1376-1380.

21. Kumar V, Rani A, Goyal L, Pratap D, Billore SD, et al. (2011) Evaluation of vegetable- type soybean for sucrose, taste-related amino acids, and isoflavones contents. International Journal of food properties 14:1142-1151.

22. Song J, Liu C, Li D, Gu Z (2013) Evaluation of sugar, free amino acid, and organic acid compositions of different varieties of vegetable soybean (Glycine max [L.] Merr). Industrial Crops and Products 50: 743-749.

9
CHAPTER

PRODUCTION POTENTIAL OF FABA BEAN (VICIA FABA L.) GENOTYPES IN RELATION TO PLANT DENSITIES AND PHOSPHORUS NUTRITION ON VERTISOLS OF CENTRAL HIGHLANDS OF WEST SHOWA ZONE, ETHIOPIA, EAST AFRICA

Tekle Edossa Kubure, Cherukuri V Raghavaiah and Ibrahim Hamza

Department of Plant Sciences, College of Agriculture and Veterinary Sciences, PO Box 19, Ambo University, Ambo, West Shoa Zone, Ethiopia

ABSTRACT

Faba bean is an important grain legume grown on vertisols of central high lands in Ethiopia, and is constrained by low yielding varieties, soil acidity, besides scanty information on optimum plant density and phosphorus nutrition. Field experiment was therefore conducted at Ambo University research farm during 2014 rainy season with the objective to determine optimum P rate and population densities for Faba bean (Vicia faba L.) genotypes grown on vertisols under rain fed conditions. The treatments comprised three genotypes (Hachalu, Walki and Local), three spacings (30 cm × 7.5 cm, 40 cm × 5.0 cm and 60 cm × 5.0 cm) and two phosphorus levels (0 kg P205/ha and 46 kg P205/ha), which were combined factorially and laid out in a split –split plot design with three replications. The results showed that the improved genotype, Walki (3,407 kg/ha) being comparable with Hachalu (3,037 kg/ha) gave substantially greater seed yield than the local cultivar (2,833 kg/ha). Seeding at 44 plants/m2 resulted in significantly higher seed and biological yields (3,815 kg/ha and 7,894 kg/ha) than 50 plants/m2 (3,074 kg/ha and 6,570 kg/ha) and 33 plants/m2 (2,388 kg/ha and 4,696 kg/ha); although the harvest index was unaltered. Fertilization of faba bean with 46 kg P205/ha resulted in substantial increase in seed (3,531 kg/ha) and biological yields (7,172 kg/ha) over no fertilizer check (2,654 kg/ha seed and 5,602 kg/ha haulm yield). The

Citation: Kubure TE, Raghavaiah CV, Hamza I (2016) Production Potential of Faba Bean (Vicia faba L.) Genotypes in Relation to Plant Densities and Phosphorus Nutrition on Vertisols of Central Highlands of West Showa Zone, Ethiopia, East Africa. Adv Crop Sci Tech 4:214. doi:10.4172/2329-8863.1000214.

harvest index tended to improve with P nutrition (49.7) over no phosphorus (47.4). Correlations worked between yield and growth and yield components showed a significant positive relation between seed yield and plant height, leaf area/plant, leaf area index, biological yield and seed yield/plant. Biomass yield is correlated with leaf area/plant, leaf area index, and plant height. Cultivation of improved varieties of faba bean Welki and Hachalu with a plant density of 44 plants/m2 (30 cm × 7.5 cm spacing) was found to be better than the local cultivar in terms of yield and yield attributes. Phosphorus fertilizer application at 46 kg P2O5/ha improved the growth, yield and yield components of faba bean on Vertisols of high lands.

INTRODUCTION

Faba bean (Vicia faba L.) is an important pulse crop grown in the highlands (1800-3000 masl) of Ethiopia, where the soil and weather are considered to be congenial for better growth and development of the crop. The crop takes the largest share of the area under pulses production in Ethiopia. The Central Statistical Agency [1] reported that faba bean is planted to 4.34% (about 5.38 lakh ha), of the grain crop area with an annual production of about 99.17 lakh quintals, 3.94% of the total grain production and yield of 18.42 q/ha in Ethiopia. It is a crop of manifold merits in the economy of the farming communities in the highlands of Ethiopia and serves as a source of food and feed and a valuable and cheap source of protein, apart from playing a significant role in soil fertility restoration in crop rotation through fixation of atmospheric nitrogen. It is a reliable source of income to the farmers, and earns foreign exchange to the country. It is mainly produced in Tigray, Gondar, Gojjam, Wollo, Wollega, Shoa and Gamo-Gofa regions of Ethiopia.

Ethiopia's faba bean export has moved north ward since the year 2000 and the major destinations are Sudan, South Africa, Djibouti, Yemen, Russia and USA, though its share in the countries pulse export is small [2,3]. The productivity of the crop is far below the potential and is constrained by several limiting factors. Faba bean is raised by farmers at varied row spacing resulting in reduced productivity. Plant density is an important production factor that ultimately determines the yield of crop per unit area. Besides, being a legume, it needs phosphorus for better root and nodule development, which is often ignored by farmers.

The low-yield potential of the indigenous cultivars is one of the most important production constraints [4,5]. Added to this, abiotic stress like water logging have all been identified as important production constraints [4]. Besides, in Vertisols of Ethiopian highlands, phosphorous is fixed and its non-availability is a limiting factor for better crop growth and development. It is known that Phosphorous nutrition plays a prime role in growth and development of roots and its role in nodulation, dry matter production, N fixation, and protein synthesis of leguminous crops is vital [6,7], though the nutrition for N is met through rhizobial fixation of atmospheric nitrogen. Hence a balanced nutrition of legumes gains significance to reap better yields, particularly under rain fed cropping conditions, where rain fall quantum and its distribution controls the crop production system. This warrants a need to generate information on the P needs of faba bean genotypes for better expression of their genetic potential in terms of growth, development and productivity [7]. Nitrogen and Phosphorous interact closely in affecting plant maturity. Phosphorous is implicated in speeding up maturity and enhancing root-shoot growth ratio, the formation of glycol-phosphate involved in photosynthesis, respiratory metabolism, apart from being a part of nucleotides (RNA, DNA) and phospholipids of membranes and play a role in energy transfer metabolism (ATP, ADP, AMP, Pyro-phosphates) [8].

The high yield potential of faba bean has not been exploited in Ethiopia and the yield in the Southwestern Ethiopian highlands is generally low (1.3 ton ha-1. compared to 1.8 ton ha-1 of world average) [9] This is largely attributed to raising low yielding local varieties, low soil pH and low P-availability in Nitisols [10,11] However, faba bean has the capacity to mobilize soil phosphorus by secretion of acids from its rhizosphere, and is therefore of important value in low-input crop rotation systems [12]. This apart, there is scanty information on the optimum plant density to reap better harvests. Therefore, the current investigation was made to evaluate the performance of faba bean genotypes as influenced by varied plant densities and phosphorus levels, in terms of growth parameters, yield and its components and root nodulation under rain-fed vertisol conditions of Ethiopia.

MATERIALS AND METHODS

Description of Study Area

A field experiment was conducted under rain-fed conditions at Ambo University research farm during the main cropping season from July to December, 2014 on vertisol. Ambo is located in the West Shoa Zone of Oromia Regional State, Western Ethiopia, at about 115 km west of Addis Ababa, located within Coordinates: 8°59'N 37°50'E, and an altitude of 2068 m a s l. The seasonal total rain fall of the area during the crop season was 570 mm, with the average minimum and maximum temperature of 9.2°C and 27.08°C, mean relative humidity of 58.02% and a mean sun shine hours of 5.62 day-1, respectively (Figure 1). The soil was characterized by Pellic vertisol [13]. The farm land preceding the current faba bean experiment was a fallow.

Genotypes Tested

In the experiment, two improved high yielding genotypes of faba bean viz; Hachalu and Walki, which are adapted to the vertisols of the highland areas, were used. These varieties are recommended for highland vertisols of Ethiopia (Ambo, Adadi, Arsi, Robe, Sinja and etc.) with altitudes of 1900-2800 masl, having a rain fall of 700-1000 mm for planting in mid-June to early July, moderately resistant to chocolate spot and rust, released from HARC /EIAR in 2010 and 2008, respectively. The days to maturity of Hachalu and Walki are 122-156 and 133-146, respectively. The potential yields of Hachalu and Walki variety were 32-45 and 24-52 quintal ha-1 on research stations and 24- 35 and 20-42 quintal ha-1 on farmer's field, respectively [14]. These two genotypes were compared with a local variety.

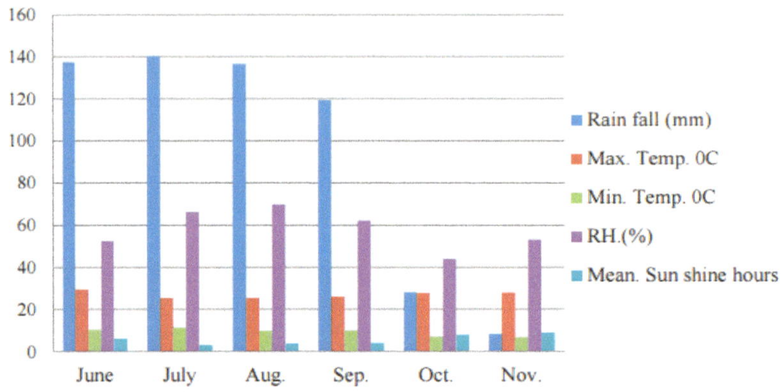

Figure 1: Meteorological data of the experimental crop season at Ambo. Source: Ambo University Meteorological Station.

Treatments and Design

The treatments comprised three faba bean genotypes (Hachalu, Walki and a local cultivar) as main- plot treatments; three spacing's (30 cm × 7.5 cm, 40 cm × 5 cm and 60 cm × 5 cm) as sub- plot treatments and two phosphorous levels (0 and 46 kg P205 ha-1) assigned to the subsub plot treatments, and were tested in split-split plot design with three replications.

Soil and Plant Analysis

Initially soil samples were collected from randomly selected sites of the experimental plots from a depth of 0–30 cm prior to cultivation and fertilizer application. The composite soil samples were analyzed for physical and chemical properties, using standard procedures for pH, CEC, organic carbon, total N, available P and K to evaluate the initial nutrient status. After the crop harvest, the soil of each treatment was analyzed for N, P, and K status. The soil physicochemical and plant tissue analysis was carried out at Holetta Agricultural Research Center (HARC), Soil and Plant Tissue Analysis Laboratory. The soil samples were air dried and ground to pass through 0.2 mm sieve for total N. Organic carbon was determined by wet digestion method as described [15]. Total N was estimated by Kjeldahl method as described[16]. Available P in soil was determined by Olsen method [17]. Soil texture was analyzed by Bouyoucos hydrometer method and Soil pH was measured by glass electrode pH meter. The plant samples (seed and haulm) were analyzed for N and P contents to calculate the nutrient uptake treatment wise. Protein content of seed was calculated based on N content of seed.

Field Operations and Crop Management

After preparatory tillage, planting was carried out on 7 July 2014. Two seeds of each genotype were planted per hill at a depth of 2–3 cm using three spacing's (30 cm × 7.5

cm, 40 cm × 5 cm and 60 cm × 5 cm) to obtain 444,444, 500,000 and 333,333 plants ha-1, respectively. Thinning was carried out two weeks after germination to maintain one plant/hill. The source of phosphorus was Di-ammonium phosphate which was applied pre planting as per treatments. Nitrogen was applied uniformly at 18 kg N ha-1 as a starter dose with urea and DAP being the sources of N. In the current experiment chocolate spot and rust diseases were observed which were managed using a fungicide Mancozeb 80 WP (Dithane M-45), at the rate of 2.5 kg a.i/ha at weekly intervals 3 times as foliar spray. The crop was harvested at physiological maturity on 19 November 2014, and subjected to sun drying to standardize the seed moisture content to 10 percent. Net plots were harvested leaving border rows to determine the per plot yields of beans and haulms.

Data Collected

Data were collected on days to emergence, days to 50% flowering, days to maturity, plant height, leaf area /plant, leaf area index, leaf number/plant; root parameters like root length, nodule number, nodule biomass; bean yield and yield components like pods/plant, seeds/pod, pod length, pod weight, 1000 seed weight, and haulm yield from five randomly selected and tagged representative plants from each net plot. Leaf area was determined by multiplying leaf length and maximum leaf breadth of all fully opened leaves on five tagged plants. It was adjusted by a correction factor of LA estimation model is LA=- 1.6923+(L*0.0161)+(W*0.0929)+(0.0062*L*W), where LA is leaf area (cm2), L is leaflet length (cm), W is the leaflet width (cm). The leaf area of faba bean was calculated using the method formulated [18]. Leaf area index (LAI) was calculated as the ratio of total leaf area of five plants-1 (cm2)/ area of land occupied by the plants.

Harvest Index (HI) was calculated as below.

Harvest index HI=Economic yield (kg/ha)/Biological yield (kg/ha) × 100

The data collected were subjected to the analysis of variance using SAS version 9.1.3 (2009), with model described below:

Model: $Y_{ijkl}=\mu+r_i+A_j+e_{ij}(a)+B_k+(AB)_{jk}+e_{ijk}(b)+C_l+(CA)_{lj}+(CB)_{lk}+(CAB)_{jkl}+e_{ijkl}(c)$

Where, μ=Population mean; r=replication; A=Main plot; ea=Main plot error; B=Sub plot; eb=Sub plot error; (AB, CA, CB, CAB)=Interaction; C=Sub-sub plot; ec=Sub-sub plot error. Wherever, the treatments showed a significant effect, the Duncan's multiple range test (DMRT) was used for means separation. The treatments were compared for their significance using calculated least significant difference (LSD) values at p=0.05.

RESULTS AND DISCUSSION

Initial Physico-Chemical Properties of the Experimental Soil

The pre- planting soil analysis showed that the texture of the soil is dominated by the clay fraction. On the basis of particle size distribution, the soil contains Sand 2.5%, Silt

22.5%, and Clay 75% (Table 1). The soil reaction (pH) of the experimental site is 6.79, which was near neutral. According to FAO, suitable pH range for most crops is between 6.5 and 7.5 in which total N availability is optimum (Table 1).

The organic carbon content of the soil was 1.17%. The soil has low organic carbon content, indicating moderate potential of the soil to supply nitrogen to plants through mineralization of organic carbon, low (0.070%) level of total N, indicating that the nutrient was not optimum for crop growth, low (5.94 ppm) available phosphorus, the K content was 1.63 meq/100 g; while the Cation exchange capacity was 1.17 meq/100 g soil [18].

Table 1: Selected physico-chemical properties of the experimental soil before sowing

Physical properties				Chemical properties					
Particle size distribution (%)									
Sand	Silt	Clay	Textural class	pH	OC (%)	CEC (Meq/100 g)	Total N (%)	Av. P (ppm)	K (Meq/100 g)
			Clay						
2.5	22.5	75		6.79	1.17	1.17	0.07	5.94	1.63

Table 2: Effect of genotypes, plant density and P levels on growth, flowering and root parameters of faba bean

Treatments	Growth parameters									Flowering		Root parameters		
	Leaf number/plant at days			Plant height (cm) at days			Basal shoots/plant	Leaf area (cm)	LAI	Days to 50% flowering	Days to maturity	No. of nodules/ plant	Nodule dry wt./ plant (g)	Tap-root length (cm)
	60	90	110	60	90	110								
Genotype														
Hachalu	19.61a	29.33a	32.17ab	70.68a	97.87a	116.94a	1.33b	54.81a	5.83a	42.28b	123a	25.28a	0.56a	18.98b
Walki	19.55a	29.11a	31.61a	71.84a	99.33a	118.33a	1.05c	52.75a	5.71a	43.00b	123a	24.28a	0.57a	18.33b
Local	19.61a	29.39a	33.00a	67.71a	93.55a	111.40a	1.61a	44.75a	4.70a	46.67a	123a	24.94a	0.55a	21.37a
Mean	**19.59**	**29.28**	**32.26**	**70.08**	**96.92**	**115.56**	**1.33**	**50.77**	**5.41**	**43.98**	**123**	**24.83**	**0.57**	**19.56**
LSD (0.05 %)	**NS**	**NS**	**NS**	**NS**	**NS**	**NS**	**0.25**	**NS**	**NS**	**3.07**	**NS**	**NS**	**NS**	**2.15**
CV%	**5.57**	**7.68**	**9.16**	**9.55**	**9.59**	**9.92**	**19.89**	**27.77**	**27.35**	**7.54**		**20**	**17.85**	**11.88**
Pl.density/m²														
44	19.56a	29.44a	32.67a	69.61a	96.44a	115.29a	1.30a	52.82a	5.79b	43.94a	123a	24.72ab	0.56a	19.82a
50	19.39a	28.89a	31.28b	70.23a	96.74a	114.69a	1.33a	47.37a	6.63a	43.94a	123a	25.44a	0.57a	19.3a
33	19.83a	29.50a	32.83a	70.39a	97.56a	116.70a	1.36a	52.12a	3.81c	44.06a	123a	24.33b	0.54a	19.56a
Mean	**19.59**	**29.28**	**32.26**	**70.08**	**96.92**	**115.56**	**1.33**	**50.77**	**5.41**	**43.98**	**123**	**24.83**	**0.57**	**19.56**
LSD (0.05 %)	**NS**	**NS**	**0.99**	**NS**	**NS**	**NS**	**NS**	**NS**	**0.74**	**NS**	**NS**	**0.96**	**NS**	**NS**
CV%	**5.26**	**5.32**	**4.23**	**2.8**	**2.72**	**3.06**	**13.02**	**17.15**	**18.74**	**0.6**		**5.33**	**17.85**	**9.55**
P_20_5 (kg/ha)														
0	19.3b	28.81b	31.78b	66.34b	91.64b	109.07b	1.30a	44.56b	4.78b	44.00a	123a	23.74b	0.54a	19.16a
46	19.93a	29.74a	32.74a	73.82a	102.2a	122.04a	1.36a	56.98a	6.05a	43.96a	123a	25.92a	0.58a	19.96a
Mean	**19.59**	**29.28**	**32.26**	**70.08**	**96.92**	**115.56**	**1.33**	**50.77**	**5.41**	**43.98**	**123**	**24.83**	**0.57**	**19.56**
LSD (0.05 %)	**0.45**	**0.52**	**0.76**	**2.29**	**3.04**	**3.76**	**NS**	**6.42**	**0.71**	**NS**	**NS**	**4.28**	**NS**	**NS**
CV%	**4.05**	**3.12**	**4.11**	**5.73**	**5.49**	**5.69**	**16.36**	**22.1**	**22.79**	**0.92**	**0**	**5.62**	**14.13**	**11.35**
Interaction														
G × Pl.				*	*	*								**

Differential Response of Faba Bean Genotypes to Plant Density and Phosphorus in Terms of Growth

Days to flowering: The number of days to 50% flowering of genotypes differed significantly, where the local genotype flowered late (47 days), in comparison with improved genotypes Hachalu (42 days) and Walki (43 days). This indicates that under rain fed conditions improved genotypes tend to flower earlier than traditional cultivars to complete life cycle early and escape any probable drought after anthesis, which is a desirable trait (Table 2). These results are in agreement with Gemechu et al. [19]. The days to 50% flowering was not altered either by different plant Densities or by phosphorus levels.

Days to maturity: Days to physiological maturity and grain filling period (from flowering to physiological maturity) for all the three test genotypes was non-significant and remained same at 123 days. Faba bean has indeterminate growth habit in that it flowers in phases depending up on the soil moisture. These results are in agreement with the findings of Gemechu et al. [19] reported that days to maturity and grain filling of faba bean genotypes ranged from 130-143 and 75-88 days, respectively. The days to maturity were not discernible in relation to plant densities and phosphorus nutrition (Table 2).

Plant height (cm): There was a significant interaction between genotypes and spacing on plant height at 60, 90 and 110 DAS (Table 2) where the genotype Welki with a spacing of 40 cm × 5.0 cm produced significantly taller plants at 60, 90 and 110 days after sowing than Hachalu and local varieties. At 60, 90 and 110 DAS Hachalu showed greater plant height at dense stands (30 cm × 7.5 cm) than at sparse plant stands. Welki possessed taller plants with 40 cm × 5.0 cm spacing than with 30 cm × 7.5 cm. Whereas, the local variety grew taller at sparse stand 60 cm × 5.0 cm than with dense stands of 30 cm × 7.5 cm and 40 cm × 5.0 cm. In this case, the local genotype grew better with wider spacing, while the improved genotypes responded to higher densities/ narrow spacing of 40 cm × 5.0 cm and 30 cm × 7.5 cm. Thus the taller stature of plants in dense crop stand is due to competition of plants for sun light (Table 3). The different plant densities could not bring about significant variations in plant height as measured at 60, 90 and 110 days after seeding. Application of 46 kg P205 /ha resulted in substantial enhancement in plant stature as compared with no phosphorus application at 60, 90 and 110 days after sowing. This positive response could be attributed to better root growth due to phosphorus application that facilitated better absorption of soil moisture and nutrients resulting in taller plants in comparison with no phosphorus application (Table 2). These results are in agreement with Gemechu et al. [19].

Leaf area per plant (cm): The leaf area/plant, an indicator of assimilatory surface exposed to sunlight, varied with faba bean genotype, and the improved types Hachalu and Walki being comparable spread out more leaf area than that of local cultivar. Thus the improved genotypes could have more photosynthesis that could have larger bearing on productivity in terms of grain and biological yield. The rate of leaf production increased linearly as temperature increased where water supply is non-limiting [20]. Different spacing's exerted significant influence on the leaf area of genotypes where in 30 cm × 7.5 cm, 40 cm × 5.0 cm and 60 cm × 5.0 cm possessed leaf areas of 1.8, 1.54 and 1.63 cm2 per leaf respectively, (52.82, 47.37 and 52.12 cm2 per plant, respectively). There was relatively less leaf area in higher density stand obtained with 40 cm × 5.0 cm spacing compared to other densities. This could be due to plant competition for below and above ground growth factors, like sun light, soil moisture and nutrient ions in dense stands. Application of phosphorus fertilizer showed a significant influence on the leaf area where in application of 46 kg P205/ha resulted in distinct enhancement in leaf area when compared with no Phosphorus fertilization (Table 2). This could be due to better growth and development of plants with application of phosphorus that ultimately lead to more leaf area than the unfertilized check.

Leaf area index (LAI): The genotype Hachalu (5.83) being on par with Walki (5.71) exhibited significantly greater leaf area index than the local cultivar (4.70) which in turn was at par with Walki. Thus higher LAI in improved genotypes which could lead to greater photosynthesis, is a desirable physiological trait that has the potential to enhance crop productivity; particularly under rain-fed conditions where often soil moisture is a limiting factor. The LAI of faba bean has been reported to vary from 3.5 to 7.0 until mid-June and there after they stabilized at 4-5 which supports our result [6]. Regarding the effect of plant densities on LAI, it was found that dense plant stands of 40 cm × 5.0 cm (6.63) being comparable with 30 cm × 7.5 cm (5.79) resulted in significant enhancement in LAI than sparse plant stand obtained with 60 cm × 5.0 cm spacing (3.81). This can be explained based on the fact that, though the leaf area/plant in dense stand was lower, the greater number of plants per unit land area could have contributed to more LAI; an indicator of more assimilatory surface per unit ground area. Higher densities were reported to favour early canopy development and increased light interception, and densities of 40 to 160 plants/m2 produced LAI varying between 3.5 to 7.0 [6]. Application of 46 kg P205/ha resulted in substantially higher LAI (6.84) than that obtained without phosphorus application (5.73) (Table 2). Phosphorus nutrition often could have complimentary effect on better uptake of nitrogen from soil through its effect on better root proliferation that could lead to greater shoot growth in terms of leaf area and LAI that could have beneficial effects on sourcesink relationships and greater dry matter partitioning efficiency. Faba beans have adopted to acquire P from low P soils and may indirectly make more P and K available for subsequent crops [21].

Number of leaves plant-1: There was no significant difference between genotypes in the number of leaves/plant at 60, 90, 110 DAS. In general the leaf number/plant varied from 19-33 at 60 days through 110 days after sowing. The Faba bean plant population densities did not also bring about significant variation in the number of leaves/plant. However, application of Phosphorus showed a significant difference between phosphorus treated and untreated plots (Table 2).

Number of basal shoots plant-1: The local cultivar produced significantly greater number of basal shoots (1.60) than Hachalu (1.33) and Walki (1.00). Different plant densities could not account for significant variations in the number of basal shoots/plant, though wider spacing tended to produce more shoots (1.36) than close spacings (1.30). Application of 46 kg P205/ha resulted in significant enhancement in the number of basal shoots/plant (1.40) in comparison with no phosphorus application (1.30) (Table 2).

Table 3: Interaction effect of genotype and plant density on plant height of faba bean

Plant height (cm)												
Genotype	**60 DAS**				**90 DAS**				**110 DAS**			
	44 pl/ m²	50 pl/m²	33 pl/m²	Mean	44 pl/ m²	50 pl/m²	33 pl/m²	Mean	44 pl/ m²	50 pl/m²	33 pl/m²	Mean
Hachalu	72.13	69.80	70.10	70.68	99.92	96.46	97.23	97.87	119.47	114.97	116.40	116.94
Walki	69.37	74.83	71.33	71.84	96.08	103.00	98.90	99.33	114.77	121.80	118.43	118.33
Local	67.33	66.07	69.73	67.71	93.32	90.78	96.56	93.55	111.63	107.30	115.27	111.40
Mean	69.61	70.23	70.39	70.08	96.44	96.74	97.56	96.92	115.29	114.69	116.70	
LSD (0.05)	1.75			2.35			3.16					

Root Parameters of Faba Bean as Influenced by Genotypes, Plant Density and Phosphorus

Root nodule number plant-1: The genotype Hachalu produced highest number of nodules/plant (25.3), closely followed by the local cultivar (24.9) and Walki (24.3) which had recorded the least number of nodules (Table 2). Thus the test genotypes were similar in nodulation and atmospheric nitrogen fixation, which is of greater significance for meeting the nitrogen requirement of the crop. Variability of faba bean genotypes for nutrient uptake and yield response has been reported by Balaban [22] increased root and shoot dry weight with fertilizer application [23]. A density of 50 plants/m2 resulted in more nodulation (25.4) than lower density (24.3) on vertisols under rain fed conditions. Application of phosphorus played a significant role in enhancing the root nodulation (25.9) of faba bean in comparison with no phosphorus (23.7). Although in legumes, nodulation ability is a genetic character, it is often influenced by crop nutrition, especially of phosphorus which is implicated in better growth and development of root system. These findings corroborate with the results of Asfaw [24] reported variability of faba bean genotypes in terms of number of nodules/plant (>30) and nodule dry weight/plant (>2 g) under rain-fed situations of Ethiopia. Increased nodulation and yield due to application of 50 kgP205/ha has also been reported in soils having 3.5 and 2.0 ppm P at ICARDA [25]. These findings are in consonance with those reported by Haque et al. [26] observed an increase in dry matter, nodulation, N fixation, P-uptake and protein yield of legumes. Faba bean requirement of P is reported to be high due to strong energy expenditure utilized during nodule formation and operation [21].

Nodule dry weight/plant (g): The dry weight of nodules/plant was more in improved genotypes Hachalu (0.56 g) and Welki (0.57 g) than the local cultivar (0.55 g) (Table 2). Looker also reported that root growth differs between varieties of faba bean, and both drought and water logging leads to fewer nodules on roots and hence less N fixation [27,28]. The root nodule biomass of faba bean has not been distinctly influenced by the plant densities either, which is in tandem with the nodule density per plant. Application of phosphorus has no discernible influence on the dry biomass of root nodules as compared with no phosphorus application.

Tap-root length (cm): It was observed that the local cultivar produced significantly longer tap roots (21.4 cm) than those of improved genotypes Hachalu (18.9 cm) and Walki (18.3 cm), indicating its adaptation to drought (Table 2). With regards to plant density, the variations in tap root length were not discernible. Application of Phosphorus however, tended to produce marginally longer tap root (19.9 cm) than no Phosphorus (19.1 cm) though it was statistically non-significant.

Table 4: Effect of genotypes, plant density and P levels on yield and yield components of Faba bean

Treatments	Pods/plant	Pod length (cm)	Pod weight/ plant (g)	No. of Seeds/ pod	1000 Seed wt. (g)	Seed yield (kg/ ha)	Biological yield (kg/ha)	Harvest index (%)
Genotype								
Hachalu	19.17b	6.19a	24.27a	2.89a	650.06a	3037.0ab	6361.4a	48.17a
Welki	18.61b	6.04a	23.47a	2.83a	523.89b	3407.4a	6973.5a	48.42a
Local	21.56a	4.58b	20.90a	2.94a	344.06c	2833.3b	5825.7a	49.10a
Mean	**19.78**	**5.61**	**22.88**	**2.89**	**506**	**3092.6**	**6386.9**	**48.57**
LSD (0.05 %)	**2.16**	**0.99**	**NS**	**NS**	**34.7**	**564.4**	**NS**	**NS**
CV%	**11.79**	**19.19**	**17.25**	**15.08**	**7.41**	**19.72**	**19.92**	**4.58**
Pl. density/m^2								
44	20.00a	5.61a	22.87ab	2.72b	508.89a	3814.8a	7894.2a	48.93a
50	19.50a	5.52a	21.46b	2.94a	517.33a	3074.1b	6570.2b	47.32a
33	19.83a	5.69a	24.31a	3.00a	491.78b	2368.9c	4696.3c	49.45a
Mean	**19.78**	**5.61**	**22.88**	**2.89**	**506**	**3092.6**	**6386.9**	**48.57**
LSD (0.05 %)	**NS**	**NS**	**1.75**	**0.22**	**16.93**	**167.98**	**367.29**	**NS**
CV%	**10.53**	**8.17**	**10.55**	**10.38**	**4.6**	**7.48**	**7.91**	**9.9**
P_2O_5 (kg/ha)								
0	19.33a	5.43b	21.75b	2.81a	492.16b	2654.3b	5601.7b	47.45a
46	20.22a	5.78a	24.01a	2.96a	519.82a	3530.9a	7172.1a	49.69a
Mean	**19.78**	**5.61**	**22.88**	**2.89**	**506**	**3092.6**	**6386.9**	**48.57**
LSD (0.05 %)	**NS**	**0.23**	**2.4**	**0.19**	**17.5**	**251.14**	**465.01**	**NS**
CV%	**11.45**	**7.35**	**18.35**	**11.54**	**6.05**	**14.2**	**12.73**	**11.64**
Interaction								
G × Pl						*	*	
G × P		*				*		
Pl × P								
G × Pl × P		*		*				

Productivity and its Components of Faba Bean as Influenced by Genotypes, Plant Density and Phosphorus

Pods plant-1: The local cultivar produced more number of pods (21.6) compared to the improved genotypes that produced 19.2 and 18.6 pods/plant in Hachalu and Walki, respectively. With regard to plant densities, the number of pods/plant did not differ significantly, though 30 × 7.5 cm spacing tended to produce more pods/plant. Faba bean did not respond to phosphorus application in terms of pod number/plant. The pod number/plant is a genetic character and is less influenced by the environment in terms of plant density and P nutrition (Table 4) results are in agreement with Gemechu et al. [19] reported 3 to 15 pods/plant for faba bean genotypes in Ethiopia. There was report of an increase in plant height, fresh weight, and pod number with 80 kgP205/ha [29].

Pod length (cm): Improved genotypes Hachalu (6.2 cm) and Walki (6.00) produced distinctly longer pods than the local cultivar (4.6 cm). Pod length is a heritable genetic character which has a bearing on ultimate seed yield of faba bean. Akin to the number of pods/plant, the length of pods was not substantially influenced by the different plant densities and varied marginally from 5.5 cm to 5.7 cm with the different spacings. Application of 46 kg P2O5/ha resulted in significant enhancement in the length of pods (5.8 cm) as compared with no phosphorus application (5.43 cm); thus indicating that pod length can be altered by P fertilization in faba bean. Significant interaction between genotypes and phosphorus on pod length showed that the genotype Walki produced significantly longer pods (6.49 cm) than the rest of the treatment combinations with the application of 46 kg P205/ha (Table 5).

There was also significant interaction between genotypes, plant density and phosphorus on pod length. The genotype Walki produced significantly longer pods (6.83 cm) than the rest of the treatment combinations with the application of 46 kg

P205/ha and a plant density of 44 plants/m2 (Table 6).

Pod weight/plant (g): Improved genotype Hachalu produced highest pod weight/plant (24.3 g) followed by Walki (23.5 g) and the local cultivar (20.9 g) (Table 4). The plant densities showed significant influence on pod weight/plant where wider spacing of 60 cm × 5.0 cm resulted in substantially greater pod weight/plant (24.3 g) than 30 cm × 7.5 cm (22.9 g) and 40 cm × 5.0 cm spacing (21.5 g). This indicates that pod weight/plant can be altered by plant spacing. The greater pod weight/plant recorded with low plant density could be attributed to less competition for growth resources like soil moisture, nutrients and sun light as compared to the dense stands. Faba bean exhibited significant response in terms of greater pod weight/plant with application of 46 kg P205/ha (24.0 g) compared to 21.7 g obtained with no phosphorus. This signifies the beneficial role of phosphorus in improving the pod weight of faba bean (Table 4).

Seeds pod-1: The seeds/pod did not vary significantly among the genotypes, while it tended to vary with plant density and phosphorus nutrition. Among the spacings, wider spacing tended to improve the seeds/pod (3.0) as compared with narrow spacings (2.7). On the other hand, phosphorus application tended to improve seeds/pod (3.0) when compared with no phosphorus (2.8). The number of seeds/pod varied distinctly when Hachalu fertilized with 46 kg P205/ha and sown at wider spacing of 60 cm × 5.0 cm (3.33) than when, sown at 30 cm × 7.5 cm (2.33) as evident from interaction (Table 7). Interaction effect of genotype, plant density and phosphorus levels on seed pod-1 of faba bean (Table 7) showed that local cultivar was relatively superior (2.94) to Hachalu (2.89) and Welki (2.83). With the application of phosphorus fertilizer and sowing at a spacing of 60 cm × 5.0 cm Hachalu produced more seeds/pod (3.33) than Walki [3] and local cultivar. In general, as the seeds/pod is a genetic character, it is less influenced by either management or P nutrition. These results are in agreement with Gemechu et al. [19] reported that seeds pod-1 of faba bean genotypes ranged from 2-3.

Test weight of seed (g): Among the genotypes Hachalu recorded substantially greater test seed weight (650 g) compared to Walki (524 g) and the local cultivar which recorded the least weight (344 g) (Table 4). This elucidates the greater source –sink relation of the improved genotypes than that of traditional cultivars. The plant densities have shown significant influence on the test weight of seed, where high density of 44 plants/m2 (508 g) and 50 plants/m2 (517 g) had seeds of greater weight than low density planting at 33 plants/m2 (492 g). Phosphorus fertilization at 46 kg P205/ha significantly improved the test seed weight (520 g) over no phosphorus (492 g). These results are in agreement with Gemechu et al. [19] found that 1000 seed weight of faba bean genotypes ranged from 249-553 g

Table 5: Interaction effect of Genotype with P level on Pod length (cm) of faba bean

Genotype		Phosphorus level		Mean
		0 P_2O_5 kg/ha	46 P_2O_5 kg/ha	
Hachalu		6.14	6.24	6.19
Walki		5.69	6.49	6.04
Local		4.56	4.61	4.58
Mean		5.43	5.78	
LSD (0.05)	0.29			

Table 6: Interaction effect of genotype × plant density × phosphorus levels on pod length of faba bean

Genotype	Pod length (cm)						
	0 kg P_2O_5/ha			46 kg P_2O_5/ha			
	44 plants/m²	50 plants/m²	33 plants/m²	44 plants/m²	50 plants/m²	33 plants/m²	Mean
Hachalu	6.60	5.67	6.17	5.77	6.33	6.63	6.19
Walki	5.50	5.73	5.53	6.83	6.53	6.10	6.04
Local	4.47	4.40	4.80	4.47	4.43	4.93	4.58
Mean	5.52	5.27	5.50	5.69	5.77	5.89	
LSD (0.05)		0.69					

Table 7: Interaction effect of genotypes, plant density and phosphorus levels on Seeds pod-1of faba bean

Genotype	0 kg P_2O_5			46 kg P_2O_5			Mean
	44 plants/m²	50 plants/m²	33 plants/m²	44 plants/m²	50 plants/m²	33 plants/m²	
Hachalu	3.00	2.67	3.00	2.33	3.00	3.33	2.89
Walki	2.33	3.00	2.67	3.00	3.00	3.00	2.83
Local	2.67	3.00	3.00	3.00	3.00	3.00	2.94
Mean	2.67	2.89	2.89	2.78	3.00	3.11	
LSD (0.05)	0.41						

Biological yield (kg ha-1): The biological yield followed a trend similar to that of seed yield/plot. Improved genotype Walki (6973.5 kg/ha) remaining comparable with Hachalu (6361.4 kg/ha) produced significantly higher biological yield/plot than that of local variety (5825.7 kg/ha). Higher plant densities represented by closer spacing of 30 cm × 7.5 cm (7894.2 kg/ha) and 40 cm × 5.0 cm (6570.2 kg/ha) produced superior biological yield kg/ha to that of wider spacing 60 cm × 5.0 cm (4696.3 kg/ha). Significantly greater biological yield kg/ha has been obtained with the application of 46 kgP205/ha (7172.1 kg/ha) than that obtained in no phosphorus plots (5601.7 kg/ha). Significant interaction between genotype and spacing on biological yield/ha showed that irrespective of genotype there was reduction in biological yield/ha with reduced plant density. The genotype Walki grown with a spacing of 30 cm × 7.5 cm produced significantly higher biological yield of 8,367 kg/ha than the other genotype spacing combinations. The next best is Hachalu raised with 30 cm × 7.5 cm spacing (7,915.77 kg/ha) and the local cultivar (7399.74 kg/ha) (Table 8).

Seed yield (kg ha-1): Among the faba bean genotypes, Walki (3407 kg/ha) and Hachalu (3037 kg/ha) gave significantly higher productivity than the local genotype (2833 kg/ha). The percentage yield enhancement of Walki and Hachalu over local cultivar was 20 and 7.2%, respectively. The superior performance of Walki could be attributed to more length of pods, greater pod weight/plant, higher seed weight/plant, more seeds/plant, higher test seed weight, ultimately leading to substantial enhancement in seed yield/plot. Bianchi et al. also reported that the number of seeds/pod and seed weight are most stable components and seed weight varies between cultivars and range from 0.1 g to 2.4 g/seed. Among the plant densities, seeding at 30 cm × 7.5 cm (44 plants/m2) resulted in superior seed productivity (3814.8 kg/ha) than that obtained with 40 cm × 5.0 cm (50 plants/m2) (3074.1 kg/ha) and 60 cm × 5.0 cm (33 plants/m2) (2388.9 kg/ha). Significant interaction between genotype × plant density on seed yield revealed that by and large, all the genotypes yielded maximum with 30 cm × 7.5 cm spacing, closely followed by 40 cm × 5.0 cm spacing, while their yields significantly dwindled with wider spacing of 60 cm × 5.0 cm. The genotype Walki seeded at a spacing of 30 cm × 7.5 cm surpassed (4166.7 kg/ha) the rest of the genotype × spacing combinations in seed productivity (Table 9). The next best was Hachalu grown at 30 cm × 7.5 cm (3777.8 kg/ha) in terms of productivity.

Fertilizing the crop with 46 kg P205/ha resulted in significantly greater seed yield (3531 kg/ha) than that without P. fertilizer (2654 kg/ ha) in vertisols. Application of 80 kgP205/ha has been reported to give 13 t/ha green pods of faba bean [15,19,22,29]. Based on results of 31 fertilizer trials (1967-1973) on faba bean concluded that response to phosphorus was high, increasing P from 36 to 72 kg/ha increased yield by 9.8% and 15.7% over control [14]. There was significant interaction between genotype × phosphorus on seed yield, where Walki fertilized with 46 kg P205/ha gave greater productivity (4074 kg/ha) than the rest of the combinations. The next best was Hachalu grown with 46 kg P205/ha (3407 kg/ha). Thus the new genotypes responded better to phosphorus application than the local cultivar (3111 kg/ha) (Table 10). Seed productivity, the culmination of vegetative and reproductive growth and developmental metabolic processes that have been taken place since the time of seeding through the maturity phases in the crop life cycle, is the economic product in which the farmer is interested. Seed yield of faba bean is a product of number of plants/m2, number of pod bearing nodes/plant, pods/node, seeds/pod and seed weight [30-35].

Table 8: Interaction effect of genotype with plant density on biological yield of faba bean

Genotype	Biological yield (kg/ha)			
	44 plants/m^2	50 plants/m^2	33 plants/m^2	Mean
Hachalu	7915.77	6465.6	4702.74	6361.37
Welki	8367	7630.58	4923.01	6973.53
Local	7399.74	5614.34	4463.11	5825.73
Mean	7894.17	6570.17	4696.29	
LSD (0.05)		400.85		

Table 9: Interaction effect of genotype and plant density on seed yield (kg/ha) of faba bean

Genotype	Population density			Mean
	44 plants/m^2	50 plants/m^2	33 plants/m^2	
Hachalu	3777.78	2944.44	2388.89	3037.04ab
Walki	4166.67	3555.56	2500	3407.41a
Local	3500	2722.22	2277.78	2833.33b
Mean	3814.81a	3074.07b	2388.89c	
LSD (0.05)		205.76		

Table 10: Interaction effect of genotype with phosphorus on seed yield (kg/ha) of faba bean

Genotype	Phosphorus level		
	0 P_2O_5 kg/ha	46 P_2O_5 kg/ha	Mean
Hachalu	2666.67	3407.41	3037
Walki	2740.74	4074.08	3407.4
Local	2555.55	3111.11	2833.33
Mean	2654.3	3530.9	
LSD (0.05)		198.4	

Table 11: Correlation coefficient (r) relationship between grain yields with various phenological, growth and yield components of faba bean

Variables	LFA	LFB	LFC	PHA	PHB	PHC	NTILL	LA	LAI	DF	NNOD	DwtNW	TRL	PPPL	PL	Pwt	Swt	SPP	SPPL	BMY	SYLDP	TSW	SYLDH	HI
LFA	—	0.920***	0.897***	0.343*	0.366 **	0.369**	-0.027ns	0.176ns	0.017 ns	0.057 ns	0.147ns	0.087 ns	0.035 ns	0.039ns	0.219ns	0.50 ***	0.492 **	0.292*	0.307*	0.056ns	0.087ns	0.006ns	0.087ns	0.039ns
LFB		—	0.773***	0.328*	0.349*	0.368**	0 ns	0.159ns	0.031ns	0.052ns	0.272*	0.159ns	0.048ns	0.052ns	0.184ns	0.447**	0.429**	0.160ns	0.280*	0.106ns	0.128ns	0.023ns	0.128ns	0.036ns
LFC			—	0.292*	0.328*	0.374**	0.133ns	0.089ns	-0.099ns	0.161ns	0.288*	0.270*	0.154ns	0.144ns	0.012ns	0.320*	0.332*	0.177ns	0.380**	0.008ns	0.057ns	-0.229ns	0.057ns	0.186ns
PHA				—	0.993***	0.958***	0.121 ns	0.63***	0.517***	-0.36ns	0.273*	0.341*	0.031ns	0.042ns	0.404**	0.52 ***	0.549***	0.120ns	0.291 *	0.565***	0.603***	0.320*	0.603***	-0.15ns
PHB					—	0.985***	0.128ns	0.660***	0.520***	-0.381**	0.298*	0.332*	0.032ns	0.045ns	0.425**	0.568***	0.594***	0.130ns	0.304*	0.560***	0.597***	0.329*	0.597***	-0.144ns
PHC						—	0.135ns	0.677***	0.516***	-0.396**	0.323*	0.311*	0.030ns	0.046ns	0.448**	0.620***	0.648***	0.148ns	0.317*	0.539***	0.574***	0.335*	0.574***	-0.131ns
NTILL							—	0.159ns	0.071ns	0.268ns	0.162ns	-0.075ns	0.204ns	0.205ns	-0.27*	0.037ns	0.044ns	0.106ns	0.369**	0.034ns	0.007ns	-0.225ns	0.007ns	-0.125ns
LA								—	0.772***	-0.529***	0.389**	0.065ns	-0.116ns	-0.117ns	0.503**	0.690***	0.687***	0.068ns	0.209ns	0.496**	0.525***	0.498**	0.525***	-0.136ns
LAI									—	-0.467**	0.409**	0.158ns	-0.14ns	-0.151ns	0.378**	0.405**	0.404**	-0.024ns	0.101ns	0.742***	0.708***	0.469**	0.708***	-0.427***
DF										—	-0.247ns	-0.301*	0.310*	0.315*	-0.775***	-0.387**	0.398**	0.087ns	0.356**	-0.252ns	-0.264ns	-0.815***	-0.264ns	0.045ns
NNOD											—	0.450**	-0.025ns	-0.031ns	0.188ns	0.381**	0.351**	0.054ns	0.016ns	0.239ns	0.267ns	0.184ns	0.267ns	-0.066ns
Dwt NW												—	-0.035ns	-0.052ns	0.178ns	0.080ns	0.091ns	0.032ns	0.023ns	0.194ns	0.175ns	0.074ns	0.175ns	-0.175ns
TRL													—	0.994***	-0.454**	-0.219ns	-0.194ns	-0.19ns	0.264ns	0.075ns	0.041ns	-0.402**	0.041ns	-0.204ns
PPPL														—	-0.434**	-0.204ns	-0.179ns	-0.188ns	0.270*	0.072ns	0.039ns	-0.391**	0.039ns	-0.204ns
PL															—	0.584***	0.582***	0.282*	-0.2ns	0.245ns	0.303*	0.756***	0.303*	0.116ns
Pwt																—	0.989***	0.281*	0.315*	0.258ns	0.291*	0.452**	0.291*	-0.044ns
Swt																	—	0.298*	0.318*	0.252ns	0.285*	0.446**	0.285*	-0.041ns
SPP																		—	0.264ns	-0.2ns	-0.173ns	-0.05ns	-0.173ns	0.101ns
SPPL																			—	-0.074ns	-0.048ns	-0.487**	-0.048ns	0.066ns
BMY																				—	0.959***	0.255ns	0.959***	-0.545***
SYLDP																					—	0.260ns	1***	-0.311*
TSW																						—	0.260ns	-0.057ns
SYLDH																							—	-0.311*
HI																								—

LFA, LFB, LFC=No of leaves at 60, 90, 110 DAS (Days after sowing); PA, PHB, PHC=plant height at 60, 90, 110 DAS; NTILL=Basal shoot (Tiller); LA=Leaf area; LAI=Leaf area Index; DF=50 % Flowering; DM=Days to Maturity; NNOD= No. of Nodule/plant; DwtNW=Nodule Dry wt./plant; TRL=Tap-root length; PPP= Pod/plant; PL=Pod Length; Pwt=Pod Weight/Plant; Swt=Seed wt./Plant; SPP=No. of Seed/pod; SPPL=No. of Seed/plant; BMY=Biological Yield/plot; SYLDP=Seed Yield/plot; TSW=1000 Seed wt; SYLDH=Seed Yield/ha; HI=Harvest Index

Harvest Index (HI): The harvest index, a measure of translocates partitioning efficiency, revealed that the genotypes did not differ in harvest index. Low density planting 60 × 5.0 cm (33 plants/m2) resulted in higher harvest index (49.45) over high density seeding at 30 cm × 7.5 cm (44 plants/m2) (48.93) and 40 × 5.0 cm (50 plants/m2) (47.32) which gave comparable harvest index (Table 4). The application of phosphorus tended to improve the harvest index (49.69) of faba bean when compared with no P application (47.45) though the variation was not discernible. Application of 50 kgP205 has been reported to enhance nodulation and yield of faba bean in soils having 3.5 and 2.0 ppm Pat ICARDA [25].

Correlation between Seed yield, growth and yield components: Correlations computed between growth, yield and yield components showed a significant positive relation between seed yield and plant height at different stages, leaf area/plant, leaf area index, biological yield and seed yield/plant. Biomass yield was correlated with leaf area/plant, LAI and plant height (Table 11).

CONCLUSION

From the foregoing discussion, it could be concluded that the new improved genotypes Welki and Hachalu out yielded the local traditional variety of faba bean, the percent yield enhancement being 20 and 7.2, respectively. Faba bean genotypes responded positively to phosphorus, where Welki fertilized with 46 kgP2O5/ha gave greater yield than Hachalu at the same level of phosphorus, both being superior to local variety. The genotypes exhibited differential behaviour in terms of seed yield when raised at different plant densities, where all the genotypes yielded maximum when raised with 44 pl/m2, followed by 50 pl/m2 while their yield dwindled with 33 pl/m2. All the genotypes performed better when sown at 44 pl/m2 density along with application of 46 kg P2O5/ha under rain fed vertisol conditions of central high lands of Ethiopia.

REFERENCES

1. Central Statistical Agency (CSA) (2013) Agricultural sample survey. Report on area and production for major crops (private peasant holdings, meher season). Addis Ababa, Ethiopia.
2. Amanuel G, Tanner DG, Assefa T, Duga D (1993) Observation on wheat and barley-based cropping sequences trials conducted for eight years in South-eastern Ethiopia. Paper presented at the 8th Regional Workshop for Eastern, Central and Southern Africa, Kampala, Uganda.
3. Lupwayi NZ, Kennedy AC, Chirwa RM (2011) Grain legume impacts on soil biological processes in sub-Saharan Africa. African Journal of Plant sciences 5: 1-7.
4. Asfaw T, Tesfaye G, Beyene D (1994) Genetics and breeding of faba bean. Cool-season food legumes of Ethiopia. Proc. First National cool-season food legumes review conference, Addis Ababa, Ethiopia.
5. Yohannis D, Kiros H, Yirga W (2015) Inoculation, phosphorous and zinc fertilization effects on nodulation, yield and nutrient uptake of Faba bean (Vicia faba L.) grown on calcaric cambisol of semiarid Ethiopia. J of soil science and environmental management 6: 9-15.
6. Igwilo N (1982) Nodulation and Nitrogen accumulation in field beans (Vicia faba L). J Agric Sci Camb 98: 269-288.
7. Tekle EK, Raghavaiah CV, Chavan A, Ibrahim H (2015)Effect of faba bean(Vicia faba L.) genotypes, plant densities and phosphorus on productivity, nutrient uptake, soil fertility changes and economics in central high lands of Ethiopia. Int J of Life Sciences 3: 287-305.
8. Salisbery FB, Ross CW (1992) Plant Physiology. 4th edn. Wadsworth Cengage Learning, New Delhi.
9. FAOSTAT (2010) Statistical databases and data sets of the Food and Agriculture Organization of the United Nations.

10. Agegnehu G, Ghizaw A, Sinebo W (2006) Yield performance and land-use efficiency of barley and faba bean mixed cropping in Ethiopian highlands. European Journal of Agronomy 25: 202–207.
11. Agegnehu G, Chilot Y (2009) Integrated nutrient management in faba bean and wheat on Nitisols. Research Report No. 78. EIAR, Addis Ababa, Ethiopia.
12. Nuruzzaman M, Lambers H, Bolland MDA, Veneklaas EJ (2005) Phosphorus uptake by grain legumes and subsequently grown wheat at different levels of residual phosphorus fertiliser. Austr J Agricult Res 56: 1041-1047.
13. Tesfaye Balemi T (2012) Effect of integrated use of cattle manure and inorganic fertilizers on tuber yield of potato in Ethiopia.Journal of Soil Science and Plant Nutrition 12: 257-265.
14. Ethiopian Institute of Agricultural Research (EIAR) (2011) Faba bean producing manual. Holetta Agricultural Research Center. Addis Ababa, Ethiopia.
15. Walkley A, Black CA (1934) An examination of the Degtjareff method for determining organic carbon in soils: Effect of variations in digestion conditions and of inorganic soil constituents. Soil Sci 63:251-263.
16. Jackson ML (1970) Soil chemical analysis. Prentice hall, England, New Jersey, p: 498.
17. Olsen SR, Dean LA (1965) Phosphorous. In Black CA (ed.). Methods of Soil analysis. Agronomy No. 9, Am Soc Agron Madisen, Wisconsin, USA,pp: 1044-1046.
18. Erdogan C (2012) A Leaf Area Estimation Model for Faba Bean (Vicia faba L.) Grown in the Mediterranean type of Climate. Kabul 7: 58-63.
19. Gemechu K, Mussa J, Tezera W (2006) Faba bean (Vicia faba L) Genetics and Breeding Research in Ethiopia: A Review. In: Ali K, Kenneni G, Ahmed S, Malhotra R, Beniwal S, Makkouk K, Halila MH(eds.) Food and Forage legumes of Ethiopia: Progress and prospects. Proceedings of the workshop on Food and Forage Legume, Addis Ababa, Ethiopia.
20. Dennetti MD, Auld BA (1980) The effect of position and temperature on the expansion of leaves of Vicia faba L. Annals of botany 46: 511-517.
21. Kopke U, Nemecek T (2010) Ecological services of faba bean. Field Crops Research 115: 217-233.
22. Balaban M, Sepetoğlu H (1991) Growth, nutrients uptake and grain yield in faba bean genotypes under various plant densities. J Ege Univ 26: 181-197.
23. Babiker EE, Elsheikh AG, Osman AH, El-Tinay (1995) Effect of nitrogen fixation, nitrogen fertilization and viral infection on yield, tannin and protein contents and In vitro protein digestibility of faba bean. Plant Foods Hum Nutr 47: 257-263.
24. Asfaw H, Angaw T (2006) Food and Forage legumes of Ethiopia progress and prospects. Biological nitrogen fixation research on food legumes in Ethiopia, pp: 172-175.
25. Murinda MV, Saxena MC (1983) Agronomy of faba beans, lentils and chickpeas. In: Saxena MC, Verma S (eds.) Proceedings of the International Workshop on

Faba Beans, Kabuli Chickpeas and Lentils in the 1980s. ICARDA, Aleppo, Syria,pp: 229-244.

26. Haque A (1986) Improved management of vertisols for sustainable crop-livestock production in Ethiopia 4 Nutrient Management. Plant Science Division Working Document 13. ILCA, Addis Ababa, Ethiopia.

27. Sprent JI (1972) The effects of water stress on Nitrogen fixing root nodules IV. Effects on whole plants of Vicia faba L. and Glyciene maximum. New Phytol 71: 603-611.

28. Sprent JI (1976) Water deficits and Nitrogen fixing root nodules. In Water defects and Plant growth Vol. IV. Soil water measurements plant responses and breeding for drought resistance.J Agric Sci Cambridge,pp: 103-152.

29. Davood H (2013) Phosphorus fertilizers effect on the yield and yield components of faba bean (Vicia faba L.). Annals of biological research 4: 181-184.

30. Thompson R, Taylor H (1977) Yield components and cultivars sowing date and density in faba bean. Ann Appl Biol 86: 313-320.

31. Ishag HM (1973) Physiology of seed yield in field beans (Vicia faba L.). J Agric Sci Camb 80: 191-199.

32. Hamissa MR (1973) Fertilizer requirements for Broad beans and Lentils Improvement and Production of Field crops, First FAO/SIDA Seminar for Plant Scientists from Africa and Near East, Cairo, Egypt.

33. Kay D (1979)Crop and Product DigestNo.3-Food legumes,London: Tropical products institute UK. Kluwer Academic Publishers, Dordrecht, Netherlands, pp: 26-47.

34. Salih FA, Ali AM, Elmubarak AA (1986) Effect of phosphorus application and time of harvest on the seed yield and quality of faba bean. FABIS Newsletter 15: 32-35.

35. Tekalign T (1991) Soil, plant, water, fertilizer, animal manure and compost analysis. Working Document No. 13. International Livestock Research Center for Africa, Addis Ababa, Ethiopia.

10
CHAPTER

EFFECTS OF POLYMICROBIAL BIOINOCULANT ON YIELD, QUALITY AND IN SITU DIGESTIBILITY OF SORGHUM SUDANGRASS IN SOUTH MISSISSIPPI

Rivera JD[1], Lemus RW[2] , Gipson ML[1] and Gipson RG[1]

[1]Mississippi Ag and Forestry Exp. Station.-White Sand Unit, Poplarville, MS, USA
[2]Mississippi State Univ. Dept. of Plant and Soil Sciences, Starkville MS, USA

INTRODUCTION

With rising input costs, beef cattle producers are forced to make decisions based upon the return on investment. Chemical fertilizer application is a common used method by beef cattle producers in the southeastern United States to improve pasture yield and quality [1]. Application of 80 kg N/ha to smooth bromegrass pastures resulted in an increase of crude protein by two percentage points, and resulted in greater bodyweight gain per ha [2]. However, Mosier [3] noted that application of commercial fertilizer is often in excess of plant uptake, with the plant only fully utilizing 17-50%. With recent changes regarding commodity and fertilizer prices, the practice of fertilizer application falls under closer scrutiny. One potential method to help increase the utilization of N fertilizer may be the use of bioinocoulants. Bioincoulants are microbes that will work in conjunction with plants to increase nutrient uptake and in theory, increase yield or reduce fertilizer input. In a review, Kennedy [4] determined that certain bioincoulants could be used to crop production systems to reduce N use and enhance farm output. Hein [5] examined the use of a bioincoulant on production of rice in Vietnam, and determined that the addition of bioinoculant resulted in increases in yield of grain. In addition to the increased yield, Cong and Dung [6] demonstrated that the use of bioincoulants maintained yield with decreased fertilizer. Most of the data that exist, involve vegetables, and grains, however, it is unclear to what extent this product might

Citation: Rivera JD, Lemus RW, Gipson ML, Gipson RG (2015) Effects of Polymicrobial Bioinoculant on Yield, Quality and In Situ Digestibility of Sorghum Sudangrass in South Mississippi. Agrotechnol 5: 136. doi:10.4172/2168- 9881.1000136.

enhance pasture productivity, since productivity (yield and quality) and affect animal performance [7,8]. Thus, our objective was to evaluate the effectiveness of this novel bioinoculant on DM yield, forage quality characteristics and digestibility of sorghum sudangrass.

MATERIALS AND METHODS

Site Establishment

The study was conducted on 6 acres at the White Sand Branch Unit, 6 km west of Poplarville, MS (30.80°N, 89.70°W, elevation 69 m). Soil is a Smithdale sandy loam (Fine-loamy, siliceous, subactive, thermic Typic Hapludults). Winter (Lolium multiflorum) was physically grazed down by the cow-herd at White Sand Branch Experiment Station in early May 2012. On May 21, 2012, the soil was mechanically broken up with a disk and prepared for planting. The experimental design was a randomized complete block in a 3 × 2 factorial arrangement and replicated three times. Treatments were three levels of nitrogen fertilizer (0, 34 and 67 kg N/ha, using (34-0-0), 0N, 34N and 67N, respectively) with or without the inclusion of bioinoculant (Sumagrow, BI). Sorghum-Sudan hybrid (Sorghum X drummondii; SugarGrazer II variety) was planted using a modified strip-till system (Plant-OVator, Tarver Equip. Folsom, LA) at the rate of 34 kg/ha on May 23 and 24, 2012. Additionally, at the time of planting, fertilizer treatment (Ammonium Sulfate, 21-0-0-24S) was applied with the seed as well, via a fertilizer box attachment on the drill. Seed and fertilizer treatments were applied in rows, there were nine rows approximately 23-m wide, with a 9 m buffer between each row. In this treatment layout, a block consisted of three rows. With each row receiving a different rate of N application (0, 34 and 67 N/ha). Each row was approximately 109 m in length. Within each row, each individual treatment cell was approximately 18 m × 23 m. Within each row each treatment cell was randomly assigned to either receive BI or not receive BI (Control). Based upon back calculation of seed used, fertilizer used, and area covered it was determined that actual planting rate was about 36 kg/ha and actual N application was 0, 37, and 65 kg N/ha.

Bioincoulant was applied according to manufacturer's instructions. The BI used in the study consisted of 3% humic acid, 1% microbes, and 96% inert ingredients. The microbial make up was Azorhizobium caulinodans – 1 × 109 CFU/ml, Bacillus subtilis – 1 × 109 CFU/ml, Pseudomonas fluorescens – 1 × 109 CFU/ml, Rhizobium meliloti – 2 × 109 CFU/ml, Rhizobium phaseoli – 1 × 109 CFU/ml, and Trichoderma virens – 1 × 109 CFU/ml. The initial application of BI occurred immediately after planting May 24. Upon completion of application on Block 1, it was decided to postpone spaying until the next day due to increasing afternoon winds. On May 25, 2012, the remaining BI was applied (blocks 2 and 3). In all instances BI was applied via a sprayer boom attached to an all-terrain vehicle (ATV) at the rate of 9.4 L/ha Plants were monitored throughout the month with height measurements taken every 2 wk. When it was determined that 75% of the plants or more were at least 66 cm tall, harvesting occurred (June 28, 2012). Four (9 m2) random samples were taken from each treatment plot. Forage was clipped to approximately 7.5 to10-cm stubble height using a mechanical hedge trimmer. Total forage within the sample area was collected, weighed and dried at 50°C for 72 h. Total dry matter produced was extrapolated from the harvested samples and forage mass was determined. Following sampling, all forage was mechanically cut and removed from the test plot area. On July 2, 2012, following harvest and biomass removal, BI was applied to the treatment plots. Based upon typical management practices, a second application of N was going to be applied following harvest, however constant rain events following the harvest prohibited the application of N for fear of damaging the plots. On July 26, 2012, the second harvest of plots occurred. All procedures were similar to the first harvest. All dried samples were ground through a 2-mm screen using a Wiley Mill. Ground samples were saved for subsequent NIR analysis for Crude Protein, Acid Detergent Fiber, Neutral Detergent Fiber,, TDN, Ca, and P using a FOSS 6500 (FOSS North America Inc., Eden Prairie, MN) and the hay equation of the NIRS Forage & Feed Testing Consortium (Madison, WI).

In Situ

The in situ procedure was approved by the Mississippi State University Institutional Animal Care and Use Committee. Approximately 1.5 g of ground dry sample from each test plot were weighed and sealed in individually pre-numbered and pre-weighed nylon digestion bags. Samples were done in triplicate and following sealing, bags were dried in a forced air oven for 48 h at 50°C, after which they were weighed. Samples were sorted by block and each block placed in a mesh laundry bag which was placed in the rumen of a cannulated heifer (3 heifers total). Heifers had access to warm season pastures (bahiagrass and bermuda grass). Following 72 h, samples were removed thoroughly hand washed until all residue was removed, samples were then dried for 48 h at 50°C after which they were weighed again. Dry matter disappearance was calculated as DM weigh prior to digestionDM weight of residue corrected for the weight of the bag. Statistics data were analyzed using PROC MIXED in SAS, with block being a random effect. Harvest period, N fertilization level and BI application and their interactions were considered fixed effects. Additionally, based upon the stark contrast observed in weather patterns during each harvest period, individual rainfall, ambient temperature and soil temperature were analyzed within each harvest period using a linear model. Significance was declared at P

RESULTS AND DISCUSSION

Weather

Weather data are presented in Table 1. No differences (P=0.87) were noted for ambient temperature between H1 and H2, with ambient temperature being 32.5°C and 35.6°C for H1 and H2 respectively. However, rainfall differed (P=0.003) with H1 receiving 0.08 cm of rain compared to 0.91 cm of rain observed in H2. Moreover, rain in H2 occurred in 17 of the 29 d of the second period. There was a tendency (P=0.06) for increased soil temperature in H1 compared to H2; 31.8°C and 30.5°C, respectively, which was probably due to differences in soil moisture.

Forage Dry Matter Yield

An effect of H ($P<0.01$) was noted for forage dry matter yield (Table 2). Greater yields were noted for the H2 compared to the H1. This is not surprising due to the lack of rainfall (Table 3) noted in June, followed by the increase in rainfall noted in July. Additionally, Ball [1] noted that greater yields in warm-season grasses will later in the growing season. No overall effect of N was observed ($P=0.32$); however, there was an N x H interaction ($P=0.05$), with greater response to N fertilizer observed during the first harvest. This difference was not surprising since N was applied at planting and rainfall constrains dater H1 impeded subsequent N application. Beyaert and Roy [9] sorghum-Sudangrass sudangrass responded favorably to N fertilization, with optimal use occurring at 100 kg/ha (over two applications). No effects of BI application ($P=0.46$), BI x N ($P=0.52$) or BI x H ($P=0.26$) interactions were noted on dry matter yield.

Forage Quality

A significant difference ($P<0.001$) was noted for H on all quality parameters (CP, ADF, NDF, TDN, Ca and P), with greater quality associated with H1 compared to H2 (Table 4). The additional rain (Table 5) during the second period may have resulted in a more rapid growth period as demonstrated by the increased noted in DM yield. Greater DM yield has been inversely related to forage quality. Cusicanqui and Lauer [10] determined that increasing the forage DM yield resulted in a decrease in digestibility. Moreover, Hoveland [11] determined that IVDMD and CP decreased in alfalfa decreased thorough the growing season. No differences were noted with regards to N ($P>0.05$), which might be due to the lack of rainfall associated during H1, and the lack of N application in H2. Due to the weather

Table 1: Least square means for weather data during the two harvest periods

Item	Harvest 1[a]	Harvest 2[a]	SE[b]	P-value[c]
Ambient temperature (°C)	32.5	32.6	0.49	0.87
Soil temperature (°C)	31.5	30.5	0.41	0.06
Rainfall, cm	0.08	0.91	0.15	0.0003

[a]Harvest 1 growing period was May 25-June 28; Harvest 2 growing period was June 29-July 26.
[b]Pooled standard error of treatment least square means.
[c]Probability that least square means differ.

Table 2: Effects of harvest date on forage DM yield and nutrient content of sorghum-sudangrass hybrid

Item	Harvest 1	Harvest 2	SE[a]	P-value[b]
Forage DM yield, kg/ha	2765.6	4680.4	171.1	0.0001
Acid Detergent Fiber, %	33.19	37.71	0.38	0.0001
Neutral Detergent Fiber, %	58.03	63	0.41	0.0001
Crude Protein, %	15.92	13.22	0.25	0.0001
TDN, %	63.04	59.52	0.29	0.0001
Ca, %	0.55	0.4	0.01	0.0001
P, %	0.27	0.25	0.002	0.0001
Mg, %	0.41	0.041	0.008	0.9
K, %	1.57	2.38	0.07	0.0001
In Situ Digestibility (72 h), %[c]	68.42	59.99	1	0.0001

[a]Pooled standard error of treatment least square means.
[b]Probability that least square means differ; NS=not statistically significant (P>0.10).
[c]*In situ* digestibility conducted over a 72 h period.

Table 3: Effects of nitrogen application on forage DM yield and nutrient content of sorghum-sudangrass hybrid by harvest data

Treatment[a]						
	Item	N0	N34	N67	SSE[b]	PP-value[c]
Harvest 1						
	Forage DM yield, lb/ac	1895.0[x]	2886.80[y]	2626.0[y]	264.57	0.05
	Acid Detergent Fiber, %	32.74	33.44	33.4	0.5	NS
	Neutral Detergent Fiber, %	57.38	58.23	28.49	0.32	NS
	Crude Protein, %	16.17	15.9	15.7	0.43	NS
	TDN, %	63.4	62.85	62.88	0.39	NS
	Ca, %	0.56	0.55	0.55	0.02	NS
	P, %	0.27	0.27	0.27	0.003	NS
	Mg, %	0.42	0.4	0.41	0.01	NS
	K, %	1.61	1.57	1.51	0.1	NS
	In Situ digestibility (72 h), %[d]	67.50[x]	66.42[y]	71.34[x]	1.72	0.05
Harvest 2						
	Forage DM yield, lb/ac	4329.5	4041.1	4166.11	216.2	NS
	Acid Detergent Fiber, %	37.91	37.15	38.07	0.44	NS
	Neutral Detergent Fiber, %	62.2	62.46	63.34	0.53	NS
	Crude Protein, %	13.14	13.24	13.28	0.61	NS
	TDN, %	59.37	59.97	59.25	0.35	NS
	Ca, %	0.42	0.39	0.39	0.02	NS
	P, %	0.25	0.25	0.25	0.002	NS
	Mg, %	0.41	0.4	0.041	0.01	NS
	K, %	2.39	2.4	2.36	0.09	NS
	In Situ digestibility (72 h), %[d]	59.7	61.39	58.87	1.72	NS

[a]Treatments were : N0=0 kg N/ha (ammonium nitrate) applied at planting; N34=34 kg N/ha applied at planting; or N67=67 kg N/ha applied at planting, due to weather, no follow up N was applied following first harvest.
[b]Pooled standard error of treatment least square means.
[c]Probability that least square means differ; NS=not statistically significant (P>0.10).
[d]*In situ* digestibility conducted over a 72 h period.
[x,y]Means without a common superscript differ (P<0.05).

Table 4: Effects of Sumagrow on forage DM yield, nutrient content, and in situ digestibility of sorghum-sudan hybrid by harvest data

TTreatment[a]					
	Item	BBI	Control	SE[b]	P P-value[c]
Harvest 1					
	Forage DM yield, lb/ac	2393.6	2544.9	216.02	NS
	Acid Detergent Fiber, %	33.01	33.37	0.45	NS
	Neutral Detergent Fiber, %	57.75	58.32	0.53	NS
	Crude Protein, %	16.22	15.63	0.34	NS
	TDN, %	63.18	62.91	0.36	NS
	Ca, %	0.56	0.55	0.01	NS
	P, %	0.27	0.27	0.001	NS
	Mg, %	0.41	0.41	0.01	NS
	K, %	1.57	1.56	0.1	NS
	In Situ Digestibility (72 h), %[d]	69.3	67.4	1.79	NS
Harvest 2					
	Forage DM yield, lb/ac	4091.09	4266.69	216.02	NS
	Acid Detergent Fiber, %	37.63	37.78	0.45	NS
	Neutral Detergent Fiber, %	62.71	63.29	0.53	NS
	Crude Protein, %	13.4	13.04	0.34	NS
	TDN, %	59.57	59.47	0.36	NS
	Ca, %	0.41	0.39	0.01	NS
	P, %	0.25	0.25	0.001	NS
	Mg, %	0.42	0.39	0.01	NS
	K, %	2.14	2.35	0.1	NS

[a]Treatments were: BI, bioinoculant applied at 11.2 L per ha following initial planting (1 d) and re-applied following first harvest (1 d); Control, no bioinoculant applied.
[b]Pooled standard error of treatment least square means.
[c]Probability that least square means differ; NS=not statistically significant (P>0.10).
[d]*In situ* digestibility conducted over a 72 h period.

Table 5: Effects of Sumagrow on forage DM yield, nutrient content, and in situ digestibility of sorghumsudangrass hybrid, overall data

Treatment[a]				
Item	Sumagrow	No Sumagrow	SE[b]	PP-value[c]
Forage DM yield, lb acre	3242.4	3405.8	890.7	NS
Acid Detergent Fiber, %	35.33	35.57	2.27	NS
Neutral Detergent Fiber, %	60.22	60.8	2.5	NS
Crude Protein, %	14.81	14.33	1.36	NS
TDN, %	61.38	61.19	1.78	NS
Ca, %	0.46	0.48	0.07	NS
P, %	0.26	0.26	0.01	NS
Mg, %	0.4	0.41	0.01	NS
K, %	1.98	1.96	0.41	NS
In Situ Digestibility (72h)[d]	64.97	64.43	1.05	NS

[a]Treatments were: BI, bioinoculant applied at 11.2 L per ha following initial planting (1 d) and re-applied following first harvest (1 d); Control, no bioinoculant applied.
[b]Pooled standard error of treatment least square means.
[c]Probability that least square means differ; NS=not statistically significant (P>0.10)
[d]*In situ* digestibility conducted over a 72 h period.

pattern noted in the present study, N may not have been effectively utilized thereby negating any effect. Jaleel [12] determined that yields and quality characteristics are reduced in drought stressed plants. No effects of BI (P>0.10) were noted on any quality parameters. Similarly no interactions (HxN, HxBI, BIxH, BIxN, or BIxHxN) were noted in the present study (P>0.05)

In Situ

Digestibility In situs were done in triplicate and all had coefficient of variation (CV, %) less than 4%. There was an effect of H ($P<0.001$) on in situ digestibility, with greater digestibility noted in H1. Similar to other forage quality measurements, we hypothesize that the increase in digestibility is probably due to the potential differences in maturity at the two harvests. However, a tendency for interaction was noted for harvest x nitrogen application (P=0.10). In the first harvest, plants fertilized with N67 had greater digestibility than the other two fertilizer treatments ($P<0.05$) As expected, no differences were noted during the second harvest. Additionally, an H x BI interaction tendency (P=0.09) was also detected; while there were no differences noted in the first harvest, in the second harvest, a tendency (P=0.07) in decreasing in situ digestibility was noted with application of BI. It is unclear in what manner the BI might have decreased digestibility; however, there may be no biological explanation for this.

CONCLUSIONS

The application of BI to sorghum-sudangrass did not affect DM yield, forage quality parameters, nor in situ digestibility. Nitrogen increased DM yield and improved in situ digestibility for the first 28 d; however, due to weather issues, it was not feasible to reapply N to plots following the first harvest. Therefore, it was difficult to measure any effects during the second growth phase. Moreover, no interactions were noted regarding BI and N application. Time of harvest was significant and did interact with some other variables. The environmental conditions between the two growth periods were so diverse that this difference was not surprising. Further studies are needed to determine treatment effects of BI on forage biomass and quality.

ACKNOWLEDGEMENT

This project was funded in part by a grant from Bio Soil Enhancers, Hattiesburg, MS and -the Mississippi Agricultural and Forestry Experiment Station.

REFERENCES

1. Ball DM, Hoveland CS, Lacefield GD (2002) Southern Forages (3rdedn) Potash and Phosphate Institute, Norcross GA.
2. Greenquist MA, Klopfenstein TJ, Schacht WH, Erickson GE, Vanderpol KJ, et al. (2009) Effects of nitrogen fertilization and dried distiller's grains supplementation: Forage use and performance of yearling steers. J Anim Sci 87: 3639-3646.
3. Moiser AR, Bleken MA, Chaiwanakupt P, Ellis EC, Freney JR, et al. (2002) Policy implications of human accelerated nitrogen cycling. Biogeochemistry 52: 477-516.
4. Kennedy IR, Choudhury ATMA, Kesckes ML (2004) Non-symbiotic bacterial diazotrophs in crop-farming systems: can their potential for plant growth promotion be better exploited. Soil Biology and Biochemistry 36: 1229-1244.
5. Hein NT (2008) The product Biogrow and improvements in its performance. In ACIAR Proc. Efficient nutrient use in rice production in Vietnam achieved using inoculant biofertilisers.
6. Thi Phan, Cong TD, Dung (2008) Interaction effects of BioGro with nitrogen and phosphorus on grain yield and nutrient uptake of rice in light textured soils of southern Vietnam. In ACIAR Proc. Efficient nutrient use in rice production in Vietnam achieved using inoculant biofertilisers 24-31.

7. Beck PA, Anders M, Watkins B, Gunter S, Hubbell D, et al. (2013) Improving the production, environmental and economic efficiency of the stocker cattle industry in southeastern United States. J Anim Sci 91: 2456-2466.
8. NRC (2000) Nutrient Requirements of Beef Cattle, Update 2000. Natl. Academies Press, Washington, DC.
9. Beyaert RP, Roy RC (2005) Influence of Nitrogen Fertilization on Multi-Cut Forage Sorghum-Sudangrass Yield and Nitrogen Use. J Agro 97: 1493-1501.
10. Cusicanqui JA, Lauer JG (1999) Plant density and hybrid influence on corn forage yield and quality. J Agro 91: 911-915.
11. Hoveland CS, Hill NS, Lowrey RS, Fales SL, McCormick ME, et al. (1988) Steer performance on birdsfoot trefoil and alfalfa pasture in central Georgia. J Prod Agric 1: 343-346.
12. Jaleel CA, Manivannan P, Wahid A, Farooq M, Somasundaram R (2009) Drought stress in plants: a review onmorphological characteristics and pigments composition. Int J Agric Biol 11: 100-105.

11
CHAPTER

ANALYSIS OF CLIMATE CHANGE IMPACTS AND THEIR MITIGATION STRATEGIES ON VEGETABLE SECTOR IN TROPICAL ISLANDS OF ANDAMAN AND NICOBAR ISLANDS, INDIA

Shrawan Singh[1] and Naresh Kumar Bainsla[2]

[1]Vegetable Science, Division of Horticulture and Forestry, Central Agricultural Research Institute (CARI), Port Blair- 744101, Andaman and Nicobar Islands, India

[2]Plant Breeding, Division of Field Crops, Central Agricultural Research Institute (CARI), Port Blair- 744101, Andaman and Nicobar Islands, India

ABSTRACT

The climate change is real phenomena and it will have impact on productivity and livelihood of primary farming options. Tropical Island regions are among the most vulnerable to this climatic catastrophe which affect local production of foods particularly perishable like vegetables. The phenomena will lead to a situation of high temperature-high humidity-low light with excess or deficit moisture. This change will help pest dynamic and buildup of their population but negatively affect the crop plants. Slight changes in crop factors have large impact on crop physiology and reproductive biology, resultantly larger change in their productivity and profitability. The change in climatic factors modifies genotype × environment interaction for physiological and economic traits by plants. Vegetables are succulent herbs which are more prone to both biotic and abiotic stresses. However, large germplasm and wide range of species and genetic diversity in tropical climate provide an opportunity to develop climate resilient genotypes to minimize the impact and assure economic returns to the farmers. The paper proposes breeding strategies for vegetable crops which could help in climate resilient breeding programme.

Citation: Singh S, Bainsla NK (2015) Analysis of Climate Change Impacts and their Mitigation Strategies on Vegetable Sector in Tropical Islands of Andaman and Nicobar Islands, India . J Horticulture 2:126. doi:10.4172/2376-0354.1000126

INTRODUCTION

Vegetable crop plants are herbaceous succulents and much prone to abiotic stresses. Most of them are grown in different agro-climatic situations than their evolutionary regions which make the vegetables more vulnerable to adverse climatic factors and associated losses. In tropical regions, the vegetable production always remains on mercy of environmental condition which varies with season and region [1]. The severity of environmental stress imposed on vegetable crops varies with their genotypes and other crop factors. Climate change factored rise in temperatures, reduction in irrigation water or drought situation, occurrence of frequent to prolonged flooding, occurrence of acidity or rise in salinity levels and increase in wind velocity are going to be major limiting factors in sustainable vegetable production in tropical islands [1]. These extremities will also affect microbial population in soil and root rhizosphere, soil health and soil fertility and increase soil erosion in tropics which ultimately reduce crop yield. The increase in decomposition rate of organic sources of nutrients and rapid losses of nutrients through leaching or washing out effect are major concern for vegetable nutrition [2]. Abiotic stresses appearing during vegetable production either can be the primary cause for disorders or they can influence the susceptibility of a harvest product to such disorders. Thus, appropriate changes are desired in crop plant morphology and physiological processes for increasing their adaptability and productivity in changing climatic situation. The abiotic stress also affects the post-harvest life of the vegetables and therefore, it is appropriate to prioritize among the breeding approaches for extending the postharvest life even in abiotic stress situation. This could be through enhancing stress tolerance of edible portion of vegetables through conventional breeding or up-regulating the associated genes and pathway by desirable modifications [3,4]. Such non-conventional breeding efforts are much awaited in tropical crops where climate chane associated factors will certainly reduce the critical time by escalating the speed of vegetable decay.

Further, the productivity of vegetables remains low in tropical islands [1,5] which could be due to genotypic and environmental factors or their interactions. Bray et al. [6] reported yield losses around 50% in vegetable crop primarily due to environmental stresses. In future, the climate associated stress events like high temperature, limited soil moisture and salinity stress will get magnified by climate change impacts [7]. Frequency of extreme events will affect the response of technologies including high yielding genotypes against soil health degradation or changes in disease and pest equilibriums and reproductive biology with modified microclimate.

Tropical islands are rich in plant diversity [8,9] which could be explored for identification of climate resilient genes or genotypes. Further, superior genotypes of vegetables are needed for effective utilization of integrated farming models which are relentlessly being developed for utilization of abundant light and soil moisture in tropics. Here, participatory role of local vegetable growers and knowledge of custodian communities like indigenous tribes about natural vegetable evolution system will speed-up the breeding process of vegetables for developing climate change resilient vegetable varieties. The Andaman and Nicobar Islands, India have been taken as model for analysis because, this geographically isolated region represents true maritime climate with 87% forest cover and significant human and livestock population [10] as well as vulnerability to extreme impact of natural disaster like tsunami [11]. Though, a systematic review was made by de la Peña and Hughes [12] for the improving vegetable productivity in a variable and changing climate for the world and specific attention to tropical continental regions. But, the tropical island ecosystem has difference in climatic factors and farming practices, therefore, suitable analysis of climate resilient breeding of vegetables was needed for improvement of vegetable sector in tropical islands. Thus, the present paper analysed the research strategies for breeding of climate resilient genotypes in vegetables for tropical island ecosystem.

GEOGRAPHY OF STUDY AREA

The archipelago of Andaman and Nicobar Islands is constituted of 572 islands in Bay of Bengal in Indian Ocean with geographical area of around 8249 sq km area. It starches between 6° N to 14° N on 92° E to 94° E and grouped into five groups of islands, namely North Andaman, Middle Andaman, South Andaman, Car Nicobar, Nancowry and Campbell Bay (Figure 1). It is located 1100 km from Chennai and 1300 km from Kolkoata, the nearest cities in mainland India. Land use pattern indicates major share under forest (87%) followed by revenue lands (7%), and only 6% for agriculture [10]. The local population is around 3.8 lacs scattered across the 38 inhabited islands and expect their food supply from limited cultivated land area of 50000 ha. Demographically, Andaman and Nicobar Islands have six indigenous tribes namely Nicobari, Jarawa, Onge, Shompen, Sentinal and Great Andamanese and various settler communities from mainland India and nearby countries. These people brought different food cultures which are still maintained but their interactions helped in synthesis of composite culture in life and field activities [13]. The tribal communities have their own nature associated life pattern which will be certainly affected by changing climate but settler farmers are major shareholders in local agriculture, therefore, their livelihood options also have high vulnerability to climate change impacts. Thus, impacts of abiotic stress are varying in nature and intensity from one island to other and also from farmers to farmers which desire mitigation strategies accordingly.

The islands have tropical humid climate with temperature ranges from 21 to 32°C, maximum in May and minimum in December, high mean value of annual rainfall (3100 mm) which distributed over May to December. Relative humidity remains high during rainy months (>85%) and touch low in February (<70%) [14]. Though, climatic conditions of islands remains favourable for round the year cultivation of tropical vegetables but rain pattern distinguish the crop seasons as rainy season (May to December) and dry season (January to April) [1]. The indigenous vegetables are major contributor in local diets and livelihood during rainy season in islands. The soils of islands are clay and silty in low lands while they are predominantly gravely clay loam, weak, sub angular blocky, friable slightly sticky in uplands soils. The C/ N ratio ranges from 16 to 21 [2], thus, soils of Andaman Islands support cultivation of wide range of vegetables.

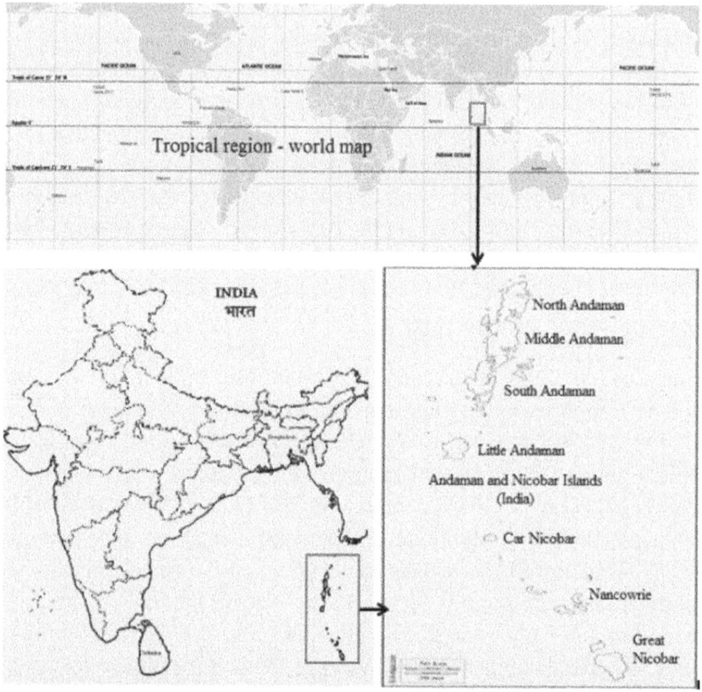

Figure 1: Tropical region of the world, India and Andaman and Nicobar Islands.

BREEDING STRATEGIES

Breeding for Water Stress

Water stress is one of the most important abiotic stresses affecting plants where excess water hampers supply of oxygen and essential nutrients to root zone and dilute the osmotic potential of cell and incite roots to secrete stress associated hormones whereas drought stress lead to plant desiccation [3]. Both excess and scarcity of water compromise the plant life and ultimately the yield. Vegetables contain moisture around 90% and any deficiency may cause serious damage to their yield and quality. The climate change impact the availability of drinking and irrigation water [14] and also challenge the vegetable sector in dry months [1]. Mitigation of the impact of flooding on vegetable sector is major challenge to vegetable growers during rainy season in tropical region where high temperature during sunny hours causes rapid wilting and death of plants.

Thus, breeding of genotypes with water stress tolerance for drought period and with high water use efficiency for drought conditions should be planned through targeted participatory approaches in natural conditions per se. For this, the local genetic resources or wild relatives of respective crops from islands can be explored for water logging or moisture stress tolerant genes. For this, the best strategy is to construct and manage the trait specific core groups from local genetic resources with regular enrichment. The islands have indigenous tribes and settler communalities which have association with such crop resources and their traditional knowledge can help in selecting the potential germplasm for targeted breeding. For example, the Solanum torvum is a naturally occurring edible plant species and offers opportunity to use in breeding against both excess and deficit of water. The physiological and genetic mechanisms in such plants should be investigated for their differential responses and use them in other crops. A targeted approach for mining good genes and their transfer in target plant using both genomics and transgenics will be more appropriate. The metabolomics and proteomics can also be used for homologus and orthologustrans-factors and alleles through substrate modification or promoter engineering. For domesticated crops, the best approach will be to screen a large number of available germplasm for moisture stress with negative mass selection method. The selected germplasm can be further utilized by pure line selection or pedigree methods along with systematic crossing programmes based upon the nature of pollination and useful heterosis.

During dry season the availability of water is limited and there are high transpiration losses and commercial species cannot withstand this stress. Therefore, genotypes with vertical leaf venation and lesser surface area, waxy leaves, green stem, stay green, shorter duration, better stomatal position, having escape mechanism can be selected. The conventional breeding approaches can be utilized for such species. However, with the genomics and transgenic approaches can be utilized for transfer of drought desirable genes like Dreb gene cassettes with universal promoters. Such genes can be sourced from local wild germplasm using model approaches.

Breeding for Tolerance to Heavy Rains

In tropical islands, the high intensity and heavy rains affect crop by severely affect physiological and reproductive processes which are associated with limited light availability, high humidity and greater susceptibility to disease and pest. Thus, breeding strategy should target multi-trait improvement and for that large scale germplasm screening followed by sequential or simultaneous breeding approach should be performed. The allele mining for suitable gene sources and transfer of such genes through conventional or non-conventional methods are options for vegetable improvement. However, due attention is required for endemic wild species of vegetable plants which have been found to be tolerant to heavy rains [9]. Thus, breeding for target stress through population improvement can be deployed for mining better alleles along with markers.

The high relative humidity (85 - 90%) affect reproductive processes like anthesis, dehiscence, pollination and fertilization in vegetable plants [12] particularly in cross pollinated crops [1]. Uptake of nutrients like calcium in tomato and sweet pepper is most affected by excess humidity. Thus, systematic efforts are required to breed the genotypes to sustain extreme humid conditions. As a matter of fact better genotypes with self-pollinated and vegetables which have vegetative portion as economic part shall be ideal choice. For this, the genetic or metabolite pathways of leafy vegetables could be investigated as they perform well in high humidity environments.

Breeding for Efficient use of Light

Cloud cover during prolonged rainy season i.e. May to December months reduce light availability to the level of a limiting factor for vegetable cultivation tropical islands, particularly of sun loving in open condition and partial sun loving in intercropping in plantations and affect their yield and quality [15]. Further, the abundant atmospheric moisture absorbs the light and makes it as a limiting factor for crop growth. The situation will increase in coming years due to climate change impact. Thus, breeding of efficient genotypes for light harvesting can be done through genetic changes or modifications in their associated traits like canopy structure, tolerance to pruning and shaping and better translocation system, vertical leaf orientation and better source to sink ratio. Further, the native leafy vegetables like Eryngium foetidum L., Ipomea aquatica L., Basella alba L., Centella asiatica Urban. and Colocasia esculenta having biomass as principal product and withstand partial to shade conditions. The indirect selection for their further refinement can be done for chlorophyll and accessory pigments. Further, vegetable in Integrated farming system (IFS) is new option for mitigating climate change impact [16], however, breeding of suitable varieties for shade tolerance, suitable for organic or less-chemical farming practices and sufficient raw materials is needed for maximing the crop returns and sustaining the system in tropical islands.

Breeding for Salinity Tolerance

Vegetable crops in coastal areas face serious threat from sea water inundation in field or irrigation streams or adverse effect of sea breeze. The natural disaster like Tsunami (2004) damaged around 8000 ha agricultural lands in Andaman and Nicobar Islands only which become unfit for vegetable cultivation [7]. The excess salinity in soil and air affect the ontogeny of the vegetable plants and reduce the productivity. In dry season, high evapo-transpiration causes substantial water loss and leave salt around the plant roots which interferes with nutrient and water uptake [3]. The plants generally try to tackle the problem by various mechanisms such osmotic stress tolerance, salt exclusion and the tolerance of tissue to accumulated salts [17]. The breeding objectives here can include selection for better salt concentration management system. The alien gene transfer from other species having adaptability to salt tolerance which can result in better survival of plants here the classical example is the transfer of saltolin rice can be used as model. Tropical islands are also blessed with rich mangrove flora out of which a few species are consumed as vegetables. The genes for salt tolerance can be mined from mangrove plant species and transferred in the commercial species. A few bacterial isolates from the rhizosphere of coastal vegetable plants from Andaman Island and sea water have been found to have excellent salt tolerance capacity (up to 15 dS-mol) (Unpublished data). Such bacterial isolates can be targeted for salt tolerance genes for improvement of vegetables like tomato.

Breeding for Temperature Extremes

Vegetables are generally sensitive to environmental extremes, and thus high temperatures and limited soil moisture are the major causes of low yields in the tropics

[9] and will be further magnified by climate change. Temperature stresses (35-45°C) which frequently occurs in mid-day hours in faulty designed polyhouses in tropical islands causes denaturation and destruct the protoplast to cell death. Such damage are observed in tomato, sweet pepper, palak, lettuce and green onion. High day temperature affect anthesis, dehiscence and fruit setting in tomato and capsicum; increase anti-nutrients and fibre content in leafy vegetables which down grade their quality; affect movement of pollinating agents and also changes stigma-pollen interaction. The rise in above-optimal temperature episodes in coming years due to climate change impact will strongly modify the reproductive processes alone or in conjunction with other environmental factors in vegetables. The size, quality, and shape of storage organs are greatly affected by day and night temperature which affects development of storage portion of tuber crops. The breeding programme for heat tolerance finds place in queue to ensure adequate supply of vegetables in coming decades.

Breeding for Mitigating Pest and Agro-Chemical Residues

The vegetable sector in tropical islands has improved a lot by technological interventions [18] but, the progress of vegetable sector has encounters serious problem of lack of resistant varieties to insect and pests [19,20] and resulted in indiscriminate use of chemical pesticides (Swarnam and Velmurgan which contributed in negative impact on health of soil and water bodies in islands [21]. Their report says that 34% vegetable samples of brinjal, okra, green chilli, crucifers, and cucurbits had pesticide residues and 15.3% samples exceeded the prescribed maximum residue limit. Thus, breeding for resistance genotypes against pest and diseases is well accepted but there is scope for finding better genotypes which can mitigate the pesticide residual effect through rapid degradation particularly post-harvest period by virtue of either inherent or induced mechanism. Further, the weed ecology is also one of the major factors for yield losses which are estimated to be very high in leafy vegetables. Use of herbicide has its own limitations in tropical ecology and breeding of weed suppressive or discoursing genotypes in vegetables are suggested. For this, the breeding of genotypes with inherent capacity of smart allelopathy for weed competitiveness is required for drastically reduction in our dependence on chemical use.

Breeding for efficient Nutrient Management

The nutrient requirement changes with crop and varieties, soil type and health, climatic situation and management practices. Micronutrients play key role in quality and yield of vegetables and islands soils are found to be deficit in Za, Ca, Cu and Bo which occurs due to washing effect in uplands and leaching out in lowlands. Further, tropical ecology favours rapid loss of chemical nutrients by soil biology or weed factors and also hasten the decomposition rate of organic sources [2,22]. Thus, use of organic agriculture may contribute in sustainability of island ecosystem but it requires suitable crop varieties while present day varieties in most of vegetables are hybrids or input responsive and not much suited for organic concept. Varietal differences for better nutrient uptake from different sources are reported in many species therefore, such genotypes are desired which can sequester higher micronutrients from soil or foliar applications.

For this, screening of available gene pools for selection or use in conventional and non-conventional breeding approaches is warranted. The breeding objective for compatibility with useful microorganism for mineralization has not been targeted yet in commercial vegetable crop species. On the other hand selective breeding approaches are required for restoration of problem soils.

Improvement of Indigenous Vegetables

Indigenous vegetables have great genetic diversity which sustains them in wide range of stresses including changing climate impact [9]. They are important component in homegardens [23] and contribute significant portion of daily diet and dietary nutrients [24] and helping in reducing micronutrient deficiencies in indigenous tribes [25]. Some of the indigenous vegetables are well adapted to stressful situation like water logging, drought and salinity. In Andaman Islands, indigenous vegetables are inexpensive, easily accessible, and highly nutritious and customarily accepted food and medicines [20]. These crops contribute around 65% of local vegetable market during rainy season and also provide dietary micronutrients and natural antioxidants to tribal and rural communities in islands. The recently initiated systematic breeding on indigenous vegetables has resulted in development high yielding genotypes/varieties of some of the indigenous vegetables (Table 1). Further, perennial vegetables like jackfruit, drumstick, Momordica etc. are well suited to fragile ecosystem [9]. However, there is need to have systematic germplasm surveys and breeding programs for identification/development of multi-stress tolerant crops/varieties to ensure the traditional diets.

Table 1: Climate change resilient improved genotypes of Indigenous vegetables

Indigenous vegetables	Variety/genotype	Characteristics
Eryngiumfoetidum L.	CARI Broad Dhaniya	Broad leaves, rich in Fe, Ca and carotenoids and high yield, tolerant to partial shade, drought and water logging conditions, suitable for organic condition
BasellaalbaL.	CARI Poi Selection	Large, dark green leaves, rich in Fe, Ca, carotenoids, ascorbic acid, high yield, tolerant to partial shade, moisture stress, and water logging conditions and disease and pests
BasellarubaL.	CARI-Poi Red	Medium size leaves with red veins, rich in Fe, Ca and anthocyanin, tolerant to partial shade, moisture stress, and water logging conditions
MomordicadioicaRoxb	CARI Kakrol	Attractive green fruits, tolerant to temporary water logged soils
AmaranthusviridisL.	CARI-AMA-Green	Green, broad leaves, early, high yield, rich in Ca, Fe and carotenoids, tolerant to moisture stress, water logging, suitable for organic cultivation
Amaranthus tricolour L.	CARI AMA-Red	Purple leaves, early, high yield, rich in Ca, Fe and anthocyanin, tolerant to moisture stress, water logging, suitable for organic cultivation
Hibiscus sabdariffaL.	CARI HS-1	Early, high yielding, rich in Fe, Ca, phenolics and carotenoids, tolerant to partial shade, moisture stress, water logging and competent to weeds and low disease and pest attack, suitable for organic cultivation
Centellaasiatica (L.)Urban	CARI CA-5	Rich in micronutrients and antioxidants, broad leaves, rapid growing, tolerant to partial shade, water logging and disease and pests, suitable for organic cultivation
IpomeaaquaticaL.	CARI NB-4	Fast growing, high yielding, rich in K, Zn, Mg, Fe, Ca, ascorbic acid, tolerant to water logging
Portulacaoleracea	CARI DB-8	Rich in Fe, Mn, Mg, phenolics and ascorbic acid, tolerant to moisture stress and water logging condition, partial shade, diseases

Breeding for Protected Cultivation

The protected cultivation technology provides favourable climatic conditions and reduces the stress levels for realizing the yield potential of crop plants. This technology has great potential in tropical islands where open cultivation of vegetables is restricted by heavy rains during rainy season [26]. Intense heat and high incidence of ultra-violet rays in dry months reduce the quality of leafy vegetables in open which otherwise very good in protected structures. Further, the land resources are very limited in islands (only 6% of geographical area is under agriculture) [10], therefore, protected cultivation has great scope to achieve self-sufficiency in vegetable sector even with less area. This is also highlighted by the Working group on Carrying Capacity of Andaman and Nicobar Islands (2009). In islands, the yield realization of improved varieties of major vegetables from mainland India are very low (2.7 to 8.6 MT/ha) in open condition. The breeding efforts through introduction and selection process identified promising varieties of different vegetables [1]. But, breeding programme in local conditions is suggested for different protected structures and for different crop seasons [27,28].

CONCLUSION

The climate change is now a real challenge to agricultural researchers working across the world. But, its real impact will be in tropical island regions where significant human population will face critical situation for food and nutritional requirements. Scientific efforts developed technologies for production management of various crops through long-term research projects but the real challenge is to provide suitable crop varieties for harnessing the potential of such management practices. The varieties with predominance of adaptive genes for different crop growing conditions could be quite strong approach. The local diversity of vegetables can contribute large in such breeding programme. The location specific breeding approaches are needed for challenging the climate change phenomena in vegetable sector.

REFERENCES

1. Singh S, Singh DR(2013a) Vegetable crops in Andaman and Nicobar Islands. Advances in Vegetable Crops.IIVR, Varanasi (In Press).
2. Pandey CB, Srivastava RC, Singh RK(2009) Soil nitrogen mineralization and microbial biomass relations; and nitrogen conservation in humid – tropics. Soil Science Society America Journal, 73: 1142-1149.
3. Taiz L,Zeiger E(2006) Plant Physiology, Fourth Edition. Sinauer Associates. Sunderland, MA.
4. Peter MA,Toivonen D, Mark Hodges(2011) Abiotic stress in harvested fruits and vegetables. Abiotic stress in plants - mechanisms and adaptations.
5. Olasantan FO(2007) Vegetable Production in Tropical Africa: Status and Strategies for Sustainable Management. Journal of Sustainable Agriculture 30: 41-70.
6. Bray EA, Bailey-Serres J, Weretilnyk E(2000) Responses to abiotic stresses. In: Gruissem W, Buchannan B, Jones R (eds.) Biochemistry and molecular biology of plants. ASPP, Rockville, MD, 1158-1249.
7. Ambast SK, Srivastava RC, Subarmani T, Chand S, Ravisankar N(2010) Design of drop spillway in undulated terrain for rainwater management in Kajunallah watershed of Andaman and Nicobar Islands. Journal of Indian Society of Coastal Agricultural Research 28: 35-39.

8. Abraham Z, Senthilkumar R, Joseph K J, Sharma TVRS et al.(2008) Collection of plant genetic resources from Andaman and Nicobar Islands. Genetic Resource and Crop Evolution 55: 1279–1289.
9. Singh S,Singh DR(2013b) Species diversity of vegetables crops in Andaman and Nicobar Islands: efforts and challenges for utilization. In: Singh D. R. Et al al. (eds.): Souvenir on Innovative technologies for conservation and sustainable utilization of island biodiversity. CARI, Port Blair 121-129.
10. Basic Statistics(2011) Andaman and Nicobar Administration.Port Blair, India.
11. Krishnan P, Dam-Roy S, George G, Srivastava RC, Anand A, et al. (2011) Elevated sea surface temperature during May 2010 induces mass bleaching of corals in the Andaman. Current Science 100: 111–117.
12. de la Peña R, Hughes J(2007) Improving vegetable productivity in a variable and changing climate. An Open Access Journal published by ICRISAT. 4: 1-22.
13. Majumdar RC(1975) Penal settlement in Andamans. New Delhi: Ministry of Education and Social Welfare, Govt of India Press, Mcgraw-Hill. pp. 256–357.
14. Srivastava RC,Ambast SK(2011) Water Policy of Andaman and Nicobar Islands. CARI, Port Blair.
15. Singh S, Singh DR, DamodaranV,JaisankarI(2010) Vegetable cultivation as intercrop in coconut plantations in Andaman and Nicobar Islands. In: Salam et al. (eds.). Proceeding Book: National Workshop on Coconut based farming system in Andaman and Nicobar islands.Directorate of Agriculture, Andaman and Nicobar Islands 54-60.
16. Ravisankar N, Pramanik SC, Jeykumar S, Singh D R, Bibi N, Shakila N,Biswas TK(2007) Study on Integrated farming System (IFS) under different resource conditions of Island ecosystem. Journal of Farming System research and Development 13: 1-9.
17. Munns R,Tester M(2008) Mechanisms of salinity tolerance.Annual Review on Plant Biology 59: 651-81.
18. Singh, S, Singh DR, Chand S, Birah A,Roy SD(2013c). Analysis of perspectives of self-sufficiency in vegetable production under tropical conditions. International Journal of Vegetable Science.
19. Jaykumar V, Prasad GS, KumarK, Sharma TVRS, Srivastava R C (2010) Hand Book on Crop Pests and Diseases of Andaman and Nicobar Islands, CARI, Port Blair. pp. 104.
20. Birah A, Sharma TVRS, Singh S, Srivastava RC (2010) Effect of aqueous leaf extract of cloves (Syzygiumaromaticum) on growth and development of tobacco caterpillar (Spodopteralitura). Indian Journal of Agricultural Sciences 80: 534-537
21. Swarnam TP,VelmuruganA(2012) Pesticide residues in vegetable samples from the Andaman Islands, India. Environment Monitoring and Assessment.

22. Pandey CB, Singh GB, Singh SK,Singh RK(2010) Soil nitrogen and microbial biomass carbon dynamics in native forests and derived agricultural land uses in a humid tropical climate of India. Plant and Soil333: 453-467.

23. Pandey CB, Lata K, Medhi R(2006) Diversity and species structure of home gardens in South Andaman. Tropical Ecology 47: 251–258.

24. Singh S, Singh DR, Salim KM, Srivastava A, Singh LB, Srivastava RC(2011) Estimation of proximate composition, micronutrients and phytochemical compounds in traditional vegetables from Andaman and Nicobar Islands. International Journal of Food Science and Nutrition 62: 765-673.

25. Rao VG, Sugunan AP, Murhekar MV, Sehgal SC(2006) Malnutrition and high childhood mortality among the Onge tribe of the Andaman and Nicobar Islands. Public Health and Nutrition 9: 19–25.

26. Arya LM, Pulver EL, vanGenuchten MT(2001) Economic Environmental, and Natural Resource Benefits of Plastic Shelters in Vegetable Production in a Humid Tropical Environment. Journal of Sustainable Agriculture17: 123-143.

27. Ceccarelli S (2009). Evolution, plant breeding and biodiversity. Journal of Agriculture and Environment for International Development 103: 131-145.

28. Leshem YY, Kuiper PJC(1996) Is there a common "GAS" (general adaptation syndrome) response to different types of physiological stress? BiologiaPlantarum, 38: 1-18.

12

CHAPTER

GIS BASED PHYSICAL LAND SUITABILITY EVALUATION FOR CROP PRODUCTION IN EASTERN ETHIOPIA: A CASE STUDY IN JELLO WATERSHED

Rediet Girma[1], Awdenegest Moges[1] and Shoeb Quraishi[2]

[1]School of Biosystems and Environmental Engineering, Hawassa University, Ethiopia

[2]School of Natural Resource and Environmental Engineering, Haramaya University, Ethiopia

INTRODUCTION

Land needs careful and appropriate use that is vital to achieve optimum productivity and to ensure environmental sustainability for future generation. This requires an effective and operative management of land information on which such decisions should be based because land is one of the non-renewable natural resource. Decision on appropriate use includes the past and present human activities [1] and the status of physical and chemical properties of the land. Land evaluation (LE) is concerned with the assessment and valuation of land when used for specified purposes. It involves the execution and interpretation of basic surveys of data on climate, soils, vegetation and other aspects of land in terms of the requirements of alternative forms of land use. To be of value in planning, the range of land uses considered has to be limited to those which are relevant within the physical, economic and social context of the area considered [2].

According to [3] the utmost pertinent solutions for the utilization of land resources in sustainable way is land-use plan by proposing alternative measures and combine the different land characteristics to solve land misuse problems. Obviously to collect, store, incorporate and analysis the different land attributes that differ spatially, Geographic Information System (GIS) could be applied [4].

Ethiopia's social and economic development is highly dependent on agriculture. Leading industry and future overall country development is also expected to be driven by the progress in the agricultural sector [5]. Even if Ethiopia is endowed with rich

Citation: Girma R, Moges A, Quraishi S (2015) GIS Based Physical Land Suitability Evaluation for Crop Production in Eastern Ethiopia: A Case Study in Jello Watershed. Agrotechnol 5:139. doi:10.4172/2168-9881.1000139.

biodiversity, throughout the country the speedy expansion of cultivation, settlements and other human activities in combination with unsustainable natural resource management even in unsuitable land has increasingly grown.

These expansions clearly exert pressure on the resource of land especially shifting of marginal and forestland in to cultivation purpose is a common practice. This is a great threat for resources as well on the resultant socio-economy, and environmental components since agriculture normally involves clearance of any natural vegetation present [2]. Due to improper land use, over exploitation and mismanagement of natural resources coupled with socio-economic factors, the problem of land degradation is on the rise and has become an issue of concern [6].

To combat land degradation, harmonizing the often-conflicting objectives of intensified human needs and socio-economic development, while maintaining and enhancing the ecology life support functions of land resources is a must. Land suitability evaluation is very important to provide information on the constraints and opportunities for the use of the land and therefore guides decisions on optimal utilizations of the resources [7]. This enables to guarantee the long-term productive potential of these resources all together by a compound effort which progressively brings the resource degradation under control [8].

At Jello watershed et al. [9] reported that an increase in cultivated and settlement lands by 55% and 107% respectively with a decline in forest lands by 80% occurred over the 30-year period since 1966. Hence, human activities are expanded onto marginal areas because the local people are entirely dependent on natural resources for their livelihood along with other physical, socio-economic and political factors. These intense changes in land use/land cover may result a significant resource imbalance due to the incompatibility of the land with land use, over exploitation of the resource and mismanagement in terms of its capability and suitability.

Though crop production is dominant in the area, the land is not evaluated/assessed and used according to its natural capability and suitability for wheat, maize and sorghum. Such types of land use practice which may seem to be highly profitable in the short run will likely to take the lion sharing to cause soil erosion and resource degradation. Such trends in agricultural production and natural resource status of the land parcel require crucial efforts for the reason that proper use of land depends on its suitability for a specific purpose that integrates different measures in sustainable way [2]. Therefore, the principal goal of this study was to perform the actual qualitative land suitability evaluation and carry out suitability mapping for the existing land use types (wheat, maize and sorghum) with an understanding of the limiting factors by integrating different information using GIS tools. Comparison between the present land use being practiced and the findings from this study was also accomplished.

MATERIALS AND METHODS

Study Area

The study area is situated in Najabas kebele of Chiro Woreda of West Hararghe Zone in the Oromiya region (Figure 1) around 326 km east of Addis Ababa. Its altitude extends between 1780-2660 m.a.s.l and the average annual rainfall is 751.3 mm [10]. During the rainy months, farmers plant sorghum, maize, wheat, inter-cropping with chat and in some parts of the area vegetables like onion, tomato and cabbage and banana as fruit tree. The area of interest covers a total of 1650 ha and agriculture is the major livelihood of the people.

Data Collection Methods

Secondary data; climatic data records (Figure 2), topographic map and the LU practices were obtained from the Department of Land Resource and Environmental Protection of the woreda.

A medium intensity soil survey (1:50,000) was used, soil sampling density of one observation for 50 ha [11]. Consequently, a total of thirty three land mapping units (LMU) were prepared over the entire 1650 ha area (Figure 3a) and one representative profile pit for each LMU was also opened (Figure 3b) and geographically referenced by using GPS. A soil sampling technique in a zigzag pattern was implemented as recommended by [12], to make it more representative; twenty subsamples of the same amount were collected from each LMU at two different fixed rooting depths (0-30 cm and 30-60 cm) separately and later, the sub-samples from similar depth were mixed carefully to made a composite sample. As a result, a total of sixty six composite soil samples over the entire area were prepared, and analyzed. Rooting depth was measured using a measuring stick; surface stoniness was estimated by selecting plots randomly to make it representative. Measurement was replicated five times and the average value was recorded in terms of areal percentage for each LMU [13]. Soil drainage class was assessed using soil profile color in combination with depth of mottling occurrence [14,15]. Flooding or inundation condition was characterized by flooding duration based on the information obtained from local people [15]. The average slope gradient was measured using clinometer aimed in the direction of the steepest slope [16].

Data Analysis Methods

Soil analysis: Bouyoucos hydrometer method was used for textural analysis and according to USDA system textural triangle was used for grouping of soil textural classes [17]. Soil pH was determined using a pH-meter with soil to water suspension ratio of 1:2.5. The OC content was determined using the standard Walkley and Black's oxidation method [12] and organic matter (OM) was computed by multiplying the organic carbon (OC) value with a constant 1.724. Electrical conductivity meter was used to measure the EC of saturated extract of the soil [12]. TN was determined by Kjeldahl standard method [18].

According to [12], the concentration of Na and K was determined by flame photometer apparatus whereas Ca and Mg were determined by atomic absorption spectrophotometric techniques. CEC was determined on the basis of displacement after washing procedure using ammonium acetate [17]. Available phosphorus was determined using Olsen's method [12,19]. CaCO3 was determined using titrimetric method with acid [20] and base saturation (BS%) was calculated by dividing the sum of extractable bases (Ca2+, Mg2+, K+, and Na+) by cation exchange capacity (CEC) and multiplying it by 100 [21,22]. Exchangeable sodium percentage (ESP) was calculated by dividing the exchangeable Na to measured CEC values and multiplying it by 100 [12,15,21].

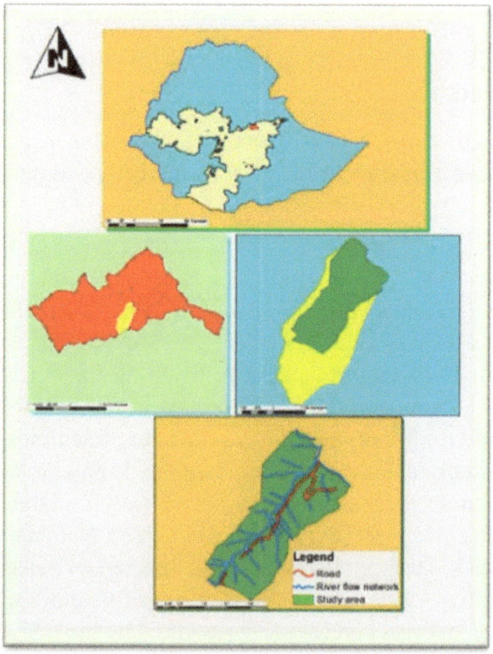

Figure 1: Location map of the study area

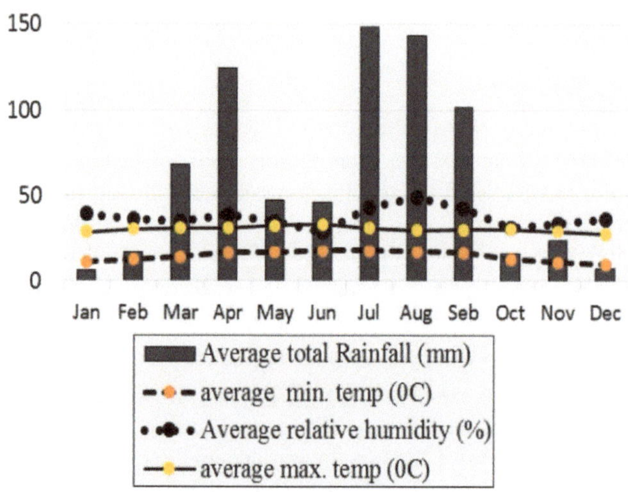

Figure 2: Climatic data.

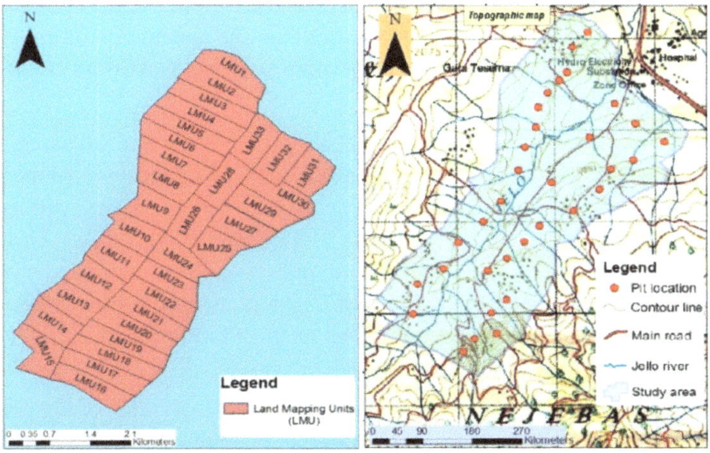

a) b)

Figure 3: LMUs (a) and pit locations (b) of the study area.

Land evaluation process: After assessing the suitability of the land for general cultivation use, the land evaluation (LE) process was proceed based on the maximum limitation method in terms of FAO's framework comparing the LCs or LQ values of each LMU with the requirements of the proposed LUT (maize, sorghum, and wheat) to identify the actual qualitative land suitability depending on physical environment data generated from topographic features, current soil characteristics, wetness condition and growing period climate data. The LE process comprised of computing the LCs values, suitability classification and land suitability mapping.

Computing the LCs values: The collected LCs data were processed and converted in to applicable LCs values (data base) using simple statistical approaches [23]. Climatic parameters during the crop growing cycle (for annual crops) was considered and an average value was calculated. In addition, the soil characteristic values changing with depth were also recalculated as depth weighted average over the 60 cm soil depth using three sections of equal thickness (20 cm) with a proportional weighting factors of 1.50, 1.00 and 0.50 from depth correction indices table [23].

Depth Correction

0-30cm: 0-20cm $\Rightarrow$ $1.5\times(20\text{-}0) = 30$

20-30cm $\Rightarrow$ $\underline{1\times(30\text{-}20) = 10}$

Sum= $\underline{40}$

30-60cm: 30-40cm $\Rightarrow$ $1\times(40\text{-}30) = 10$

40-60cm $\Rightarrow$ $\underline{0.5\times(60\text{-}40) = 10}$

Sum= $\underline{20}$

Therefore, the recalculated depth weighted average soil characteristic values over the total 60 cm soil depth was calculated by dividing the summation of the product of depth correction and soil characteristic value from 0-30 cm and 30-60 cm soil depths by the total depth (60 cm).

Suitability classification: Once the database was created and prepared, for each LCs values layers were made using GIS (Figure 4a). For the accomplishment of GIS assisted land suitability evaluation, querying analysis (attribute queries) was used based on the attributes of every LMU to generate individual land suitability classification (LSC) for each LCs values/layers separately in reference to the suggested cropspecific requirement. After merging the individual LSC layers using the GIS Merge window, the overall LSC was assigned for each LMU by its most limiting characteristics (Figure 4b).

Suitability mapping: As an output, the land suitability map of the study area was displayed and shown on individual, transparent maps using different colors to indicate the suitability classes which had all its corresponding land qualities and land characteristics in its attribute table with the help of ArcGIS (Figure 4). Comparison was accomplished between the findings from this study and what is being practiced today to select relatively the better land use option.

Results and Discussion

Diagnostic LCs with set of values is illustrated in the table below for every LMU (Table 1). The soil characteristics values explained were the recalculated depth weighted average values. The deep rooting depth of representative soil profiles was measured to be more than 100cm for all LMU, which was treated as the ideal depth for annual crops as described by [12,13].

Land Suitability Evaluation for General Cultivation

LMU32 (50 ha) is the only one assigned as highly suitable (S1); LMU1, 6, 7, 8, 18, 23 and 28 (350 ha) were grouped under moderately suitable (S2) class; nineteen (950ha) LMUs were classified as marginally suitable (S3) class and LMU2, 9, 14, 16, 17 and 19 (300 ha) were not suitable (N) due to fertility status and slope condition for cultivation purpose in general.

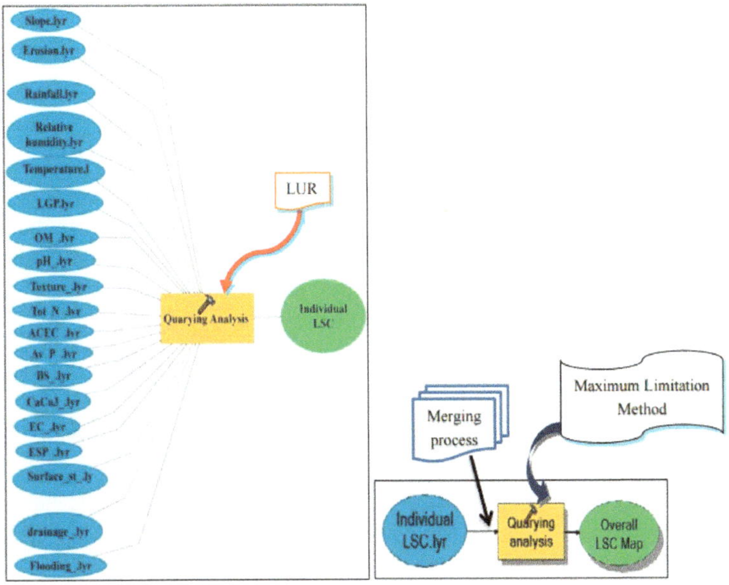

a) b)

Figure 4: (a) GIS model used for individual LSC and (b) overall LSC

Table 1: The recalculated depth weighted average soil characteristics values

LMU	Soil texture class	ESP %	EC dSm^{-1}	$CaCO_3$ %	CEC Cmol/kg	OM %	Av. P ppm	TN %	pH H_2O	BS %	Slope %	Drainage class	Flooding risk	Surface stoniness, %
1	SCL	5.9	0.1	5.4	57	2.49	1.75	0.13	7.5	75.2	5-8	Well	Nil	3
2	Sandy loam	7.72	0.1	5.3	47.6	0.45	0.49	0.04	7.1	82.4	8-15	Well	Nil	3.5
3	SCL	7.27	0.25	7.4	56.9	1.15	0.88	0.11	7.23	71.3	3-8	Well	Nil	12
4	SCL	7	0.17	4.8	49	0.65	1.27	0.05	6.93	79.9	3-8	Well	Nil	4.2
5	SCL	6.9	0.09	3.4	49.6	1.51	1.09	0.08	7.07	87.4	3-8	Well	Nil	10
6	Loam	10.29	0.18	5.13	43.5	1.99	3.3	0.14	7.73	87.6	3-8	Moderate	Nil	2.1
7	SCL	7.74	0.14	2.97	43.9	2.65	1.64	0.07	7.47	70.2	8-15	Well	Nil	2
8	SCL	6.16	0.26	6.6	56.5	3.32	2.56	0.12	7.33	83.1	3-8	Well	Nil	0.2
9	Sandy loam	8.34	0.14	3.87	41.5	0.53	0.5	0.05	6.97	71.8	5-8	Well	Nil	1
10	Sandy loam	6.73	0.11	7.7	45	2.22	1.83	0.13	6.97	92.1	3-5	Well	Nil	3.4
11	SCL	6.17	0.21	4	56.9	2	2.27	0.2	7.5	45.2	5-8	Well	Nil	4
12	SCL	6.48	0.15	3.87	52.2	1.96	1.56	0.12	7.27	64.8	8-15	Well	Nil	5.7
13	Loam	5.35	0.28	6.6	54	3.59	7.19	0.19	7.17	64.9	3-8	Moderate	Slight	8.2
14	SCL	9.65	0.14	4.47	46	1.14	3.52	0.08	6.93	77.7	>15	Well	Slight	9
15	SCL	12.38	0.21	8.4	41.8	3.38	4.22	0.15	7.1	81.2	3-5	Well	Nil	4.6
16	SCL	6.97	0.15	4.9	49.7	3.15	7.17	0.14	7.07	72.1	>15	Well	Nil	2
17	Clay loam	6.24	0.21	8	61.3	1.27	1.99	0.07	6.97	82.8	>15	Moderate	Nil	2.5
18	SCL	6.48	0.29	6.3	53.6	2.85	8.39	0.13	6.77	86.2	8-12	Well	Nil	3
19	Loam	4.61	0.14	9.03	61.8	3.59	5.81	0.13	7.3	78.9	>15	Moderate	Nil	2.9
20	Loam	7.09	0.16	3.6	47.8	2.31	7.07	0.12	7.57	78.1	8-15	Moderate	Nil	4.8
21	SCL	6.3	0.12	9.53	57.6	2.54	4.33	0.53	7.13	84.3	8-15	Well	Nil	5.2
22	SCL	8.13	0.07	11.27	43.1	2.87	4.8	0.14	7.73	89.9	3-5	Well	Nil	6
23	SCL	7.74	0.2	4.8	52.9	2.53	7.12	0.77	8.13	75.9	3-5	Well	Nil	2.8
24	SCL	6.05	0.1	4	43.6	2.93	7.33	0.59	6.93	88.9	3-8	Well	Nil	7.4
25	Clay loam	7.64	0.24	7.13	50.4	3.57	5.9	1.32	7.73	90.8	8-10	Moderate	Nil	4
26	Loam	6.21	0.14	6.5	53.5	3	5.9	1.13	8.13	69.1	5-8	Moderate	Nil	4.5
27	Clay loam	8.61	0.16	4.6	42.2	2.71	4.73	0.67	7.93	84.8	8-10	Moderate	Nil	12
28	SCL	8.07	0.12	6.1	52.1	2.87	5.83	0.54	7.37	83.8	3-8	Well	Nil	2.5
29	SCL	12.28	0.1	5.9	39.7	3.47	7.47	0.83	7.73	90.6	8-15	Well	Nil	4
30	SCL	6.69	0.12	4.8	52.6	3.34	6.9	0.67	7.8	89.7	3-5	Well	Nil	3.6
31	SCL	6.04	0.1	9.4	50.4	2.34	2	0.84	7.07	70	5-8	Well	Nil	6.2
32	SCL	5.71	0.12	6.9	49.4	2.98	3.2	1.31	7.04	88.6	0-3	Well	Nil	2.9
33	Clay loam	7.8	0.11	6.2	52.8	3.25	4.07	0.86	6.7	74.9	0-3	Moderate	Nil	3.4

Remark: "SCL"= Sandy clay loam

Individual LSC for each LCs

On the bases of individual LCs values, a separate class was rated (except for the rooting depth />100cm/ and temperature assigned as S1 concerning the three LUT; rainfall labeled as S1 for sorghum, S2 for wheat and maize; relative humidity was allocated as S2 for maize and sorghum).

Suitability ratings for wheat: Individual LCs were examined for wheat, accordingly the suitability percentage are shown in Figure 5. Suitability ratings for maize: In a similar way, individual LCs values were also matched with the requirement of maize. The suitability percentage is shown in Figure 6. Suitability ratings for sorghum: Individual LCs suitability percentage is also shown in Figure 7.

Overall Land Suitability Classification

In general, land suitability classification of the mapping units centered on the most limiting land characteristics was classified and labeled into different suitability classes using the GIS quary builder technique (Figure 8). The overall suitability map for each LUTs was presented as an output after merging the individual LSC layers acquired. Overall land suitability classification for wheat: The overall suitability class of the study area for wheat cultivation was generally grouped into three ratings (Figure 9a) as moderately suitable (S2); marginally suitable (S3); and not suitable (N). Soil fertility, topographic feature, surface stoniness in conjunction with rainfall (LMU23 and 30) and pH (for LMU23) were considered as the limiting factors in general.

Overall land suitability classification for maize: The interpretation from the overall suitability map (Figure 9b) generated with the help of GIS tools shows that, no LMU was assigned as moderately suitable for maize production. The restrictive factors inducing the two suitability class (S3 and N) assigned were owed by the soil fertility status, pH and topographic factors.

Overall land suitability classification for sorghum: Similarly, the overall suitability class for sorghum cultivation falls under marginally suitable (S3) and unsuitable (N) (Figure 9c). The suitability class was brought by the dominant limiting factors as soil fertility, topographic condition, RH, together with textural class (for LMU2, 9 and 10), surface stoniness aimed at LMU13 and pH for LMU23 and 26. Concerning the three land utilization types, non-suitability class (N) was also observed as a mutual rating for 800ha (48% or LMU1, 2, 3, 4, 5, 7, 8, 9, 10, 11, 12, 14, 16, 17, 19 and 31). LMU25 and 27 were assembled to be marginally suitable for maize and wheat. For that of maize and sorghum, there was also 100 ha (LMU23 and 30) grouped under marginal suitability (S3). As far as the suitability map of wheat and sorghum was referred;

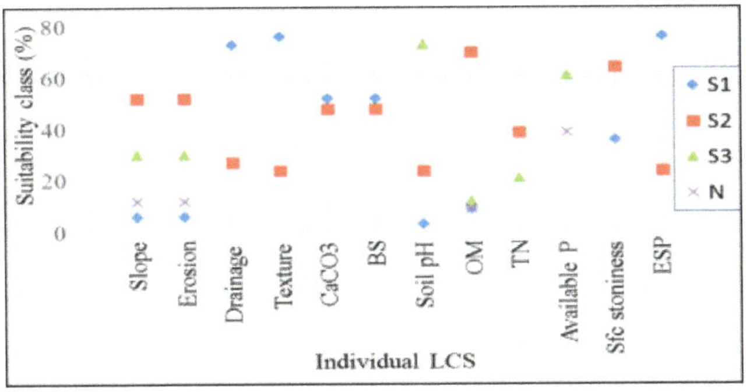

Figure 5: Individual LSC percentage for wheat

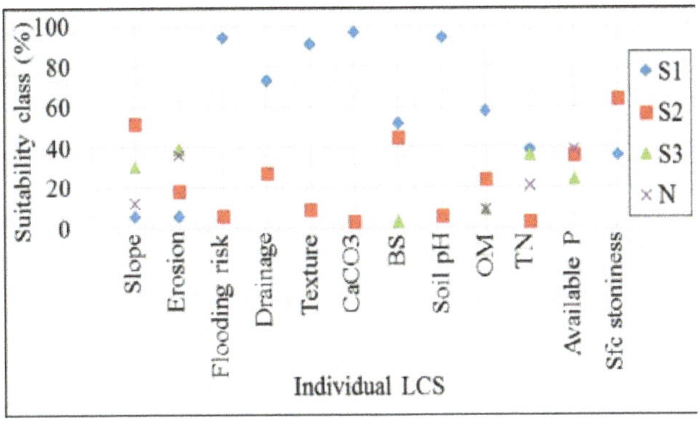

Figure 6: Individual LSC percentage for maize

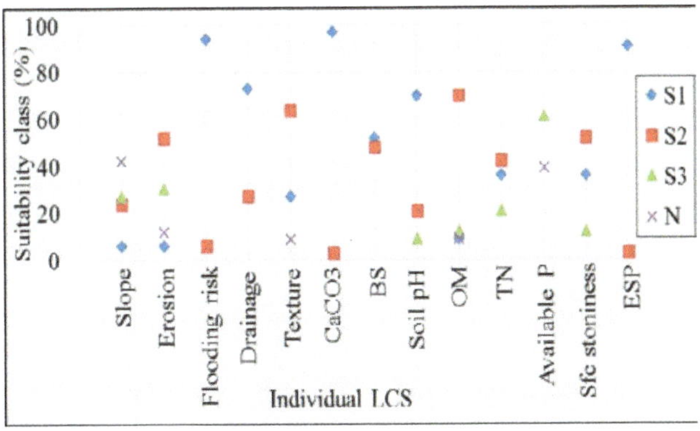

Figure 7: Individual LSC percentage for sorghum

LMU18, 20, 21 and 29 (200 ha) was unsuitable (N) for both LUTs. The outcome of this study also tells that none of the thirty three LMU falls under highly suitable (S1) class.

Currently, farmers on the entire watershed cultivate maize and sorghum (more dominantly) intercropping with chat. According to the comparison made between the existing land use being practiced and the findings from this study, 800 ha (48%) and 1100 ha (67%) area of land was mismatched (currently not suitable) for maize and sorghum cultivation respectively.

Based on the analysis, wheat cultivation is relatively better (moderately suited) than the land use being practiced (maize and sorghum) on the bases of the present situation for 100 ha (LMU23 and 30). Comparatively maize cultivation is the other option (it is marginally suited) for LMU18, 20, 21 and 29 (200ha) and wheat or maize is better on LMU25 and 27 (100 ha) rather than sorghum cultivation at present. On the other hand, none of the three land utilization types are suitable for 800ha or 48% of the total area.

The continuing of existing land use (LU) practices beyond the natural ability of the land had been considered as a catalytic agent for the exponential depletion of the present scarce soil resource, the low soil fertility status for instance. The nonstop expansion of agricultural practice was the driving force for the decline of forest area and in some parts of the surveyed area long and steep slope is used for cultivation purposes.

Concurrently unless and other wise measures are taken, these augment soil erosion, resource degradation and adverse changes in river regimes (for instance, frequent flooding and absence of dry season stream flow) leading to irreversible degradation without hyperbole.

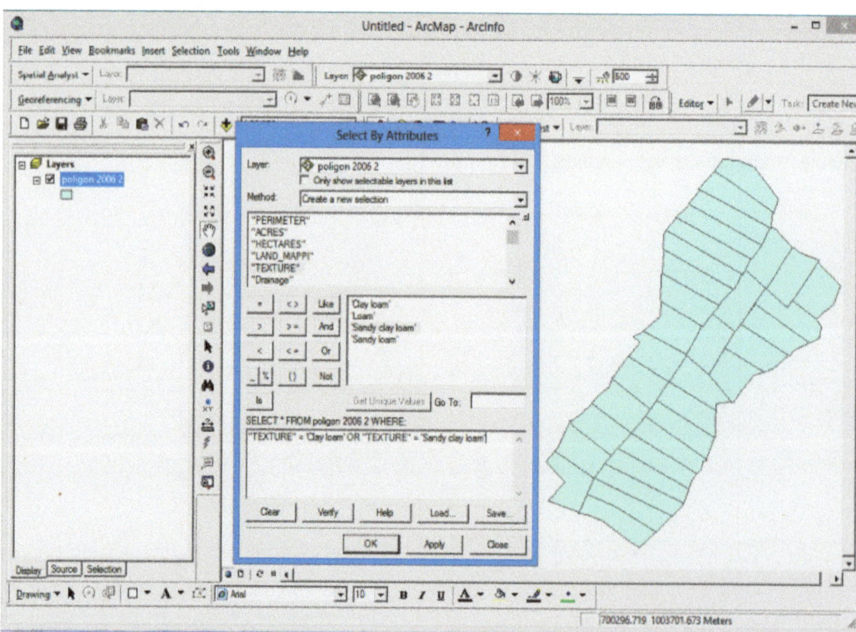

Figure 8: Quarying analysis window used.

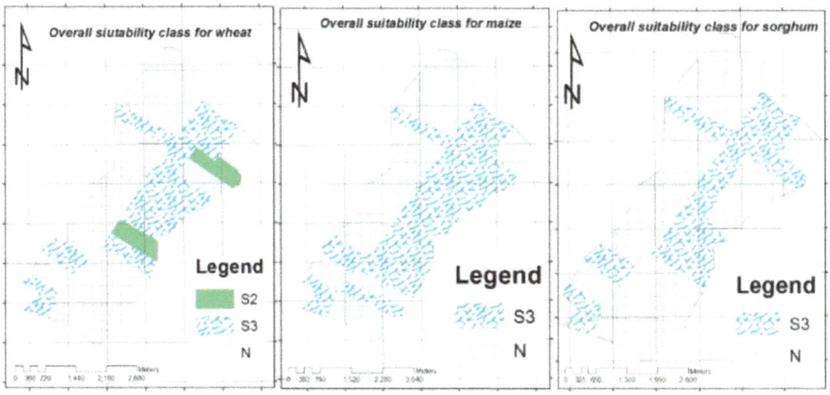

(a) (b) (c)

Figure 9: (a) Overall land suitability map for wheat, (b) maize and (c) sorghum

CONCLUSION

Currently, the surveyed area was assembled into three suitability classes as moderately suitable (S2) for 6%, marginally suitable (S3) for 33% and 61% unsuitable (N) for wheat. On the other hand two suitability classes were observed for maize (52% and 48%) and sorghum (33% and 67%) cultivation as marginally suitable (S3) and unsuitable (N) respectively. From the overall suitability ratings attained, presently none of the thirty three land mapping units fell under highly suitable (S1) class to any of the three land utilization types. Based on the individual LCs, exceedingly available P (named as the most severe limiting factor) of the soil was the only one responsible for not to be characterized as high suitability (S1). Therefore, major limiting land characteristic for crop production in the district had been the low fertility status of the soil attributed to the low amount of available P. This also indicates that such lands can be degraded and easily loose the productive potential if the existing land use practices are ongoing and no well-timed appropriate measures are undertaken.

ACKNOWLEDGEMENTS

We are grateful for the support of the Federal Ministry of Agriculture. Land users and experts from bureau of Land Resource Management of Chiro woreda provided me information and participated in data collection are greatly acknowledged.

REFERENCES

1. KeshavarziA, Sarmadian F,Ahmadi A (2011)Spatially-based model of land suitability analysis using Block Kriging. Australian Journal of Crop Science 5:1533-1541.
2. FAO (1976) A framework for land evaluation.FAO soil bulletin 32, Rome.

3. Laosuwan T, Sangpradid S, Chunpang P (2013) Suitable areas for economic crops based on GIS and physical land evaluation model. International Journal of Soft Computing and Engineering (IJSCE) 3.
4. Mahmoud A, Shendi M.M, Pradhan B, Attia F (2009)Utilization of remote sensing data and GIS tools for land use sustainability analysis: case study in El-Hammam area, Egypt.Central European Journal of Geosciences 1: 347-367.
5. TeshomeYibarek, KibebewKibret, HelufGebrekidan, ShelemeBeyene (2013) Physical land suitability evaluation for rain fed production of cotton, maize, upland rice and sorghum in Abobo Area, western Ethiopia. American Journal of Research Communication1: 296-318.
6. Panhalkar S (2011) Land capability classification for integrated watershed development by applying remote sensing and GIS techniques. Journal of Agricultural and Biological Science. Asian Research Publishing Network (ARPN) 6: 4.
7. Rabia AH (2012) A GIS based land suitability assessment for agricultural planning in KilteAwulaelodistrict, Ethiopia. The 4th International Congress of ECSSS, EUROSOIL, Bari, Italy.
8. Mitiku H, Herweg K, Stillhardt B (2006) Sustainable Land Management-A New Approach to Soil and Water Conservation in Ethiopia.
9. Mohammed A (2006) Land use/cover dynamics over a period of three decades in the Jello micro-catchment, Chercher highlands, Ethiopia. Ethiopian Journal of Natural Resources.
10. BoA (2001) Result oriented community based participatory watershed development plan in Jello. Oromiya region, West Hararghe zone, Chiroworeda.
11. Deckers J, Spaargaren O, Dondeyne S (2002) Soil survey as a basis for land evaluation. Land Use, Land Cover and Soil Sciences. II: Encyclopedia of Life Support System.
12. FAO (2007) Methods of analysis for soils of arid and semi-arid regions. Rome, Italy.
13. USDA-NRCS (2008) Soil Survey Manual (Complete Print Friendly Version) NRCS Soils.
14. Briggs RD (1994) Site classification field guide. Cooperative Forestry Research Unit Technical Note 6. Miscellaneous publication 724.
15. Ritung S, WahyuntoAgus F, Hidayat H (2007) Land Suitability Evaluation with a case map of Aceh Barat District. Indonesian Soil Research Institute and World Agroforestry Centre, Bogor, Indonesia.
16. FAO (2006) Guidelines for soil description (4th edition), Rome.
17. Soil Survey Staff (2009) Soil Survey Field and Laboratory Methods Manual. Soil Survey Investigations Report No. 51, Version 1.0. R. Burt (ed.) U S. Department of Agriculture, Natural Resources Conservation Service.

18. SahlemedhinSertsu, TayeBekele (2000) Procedures for soil and plant analysis. National Soil Research center. Ethiopian Agricultural Research Organization. Technical paper No 24.

19. TayeBelachew, YifruAbera (2010) Assessment of soil fertility status with depth in wheat growing highlands of Southeast Ethiopia. World Journal of Agricultural Sciences 6: 525-531.

20. USSLS (1954) Diagnosis and improvement of saline and alkali soil. Handbook 60, US Government Printing Office, Washington DC.

21. Moore G (2001) Soil guide: A handbook for understanding and managing agricultural soils. Agriculture Western Australia Bulletin No 4343.

22. USDA-NRCS (2004) Soil Survey Laboratory Methods Manual. Soil Survey Investigations Report No. 42 Version 4.0.

23. Van Ranst E, Ann verdoodt (2005) Laboratory of Soil Science. Land evaluation Part II: Qualitative Methods in Land Evaluation. International center for physical land resources. Belgium.

13

CHAPTER

PATH COEFFICIENT ANALYSIS FOR SOME CHARACTERS ON FRUIT AND SEED YIELDS OF OKRA

Sameh AM Abd-allah

Horticulture Research Institute, a.r.c., Egypt

INTRODUCTION

Okra is a multipurpose use crop with tender and delicious pods. In West Africa; leaves, buds and flowers of okra are, also, consumed. The dried seeds provide oil, protein, vegetable curd and a coffee additives or substitutes. Okra dry seeds are reported to contain 18-20% oil and 20-25% crude protein [1]. Foliage can be used for biomass and the dried stems serve as a source of paper pulp or fuel. Okra, to a certain extent, is used in canned, dehydrated or frozen forms. It has an average nutritive value of 3.21 that is higher than tomato, eggplant and most cucurbits except bitter gourd, so, it may be considered as one of the important vegetable crops in the tropical and subtropical regions of the world; especially in Egypt at the summer season [2].

It is necessary to determine the factors or traits that influence fruit and seed yields of okra, directly and indirectly or both; hence, their response to selection can be predicted. As genetic correlation is the correlation of breeding value and results mainly from pleiotropy and dealing with the assumption simply measures mutual relationships without presumption of causation, so that in path-coefficient analysis, the researcher specifies the plausible causes and measures their relative importance as independent variables [3,4]. Moreover, Agriyo et al. [5] calculated the correlation coefficients of 30 okra-genotypes and found that coefficients varied between seasons; For instance, pod yield/plant was: 1) genotypically correlated with edible pod weight, length and width, mature pod length number of branches/plant, 2) phenotypically correlated with number of branches/plant and edible pod length, and 3) environmentally correlated with edible pod weight and length, and final plant height.

Citation: Abd-allah SAM (2015) Path Coefficient Analysis for Some Characters on Fruit and Seed Yields of Okra. J Horticulture 2:140. doi: 10.4172/2376-0354.1000140.

The present study was directed to estimate the direct and indirect effects of some characters related to fruit and seed production in some okra varieties and lines, originated from a breeding program at Sabaheia Horticultural Research Station.

MATERIALS AND METHODS

Two field experiments were conducted during 2013 and 2014 summer seasons at Sabahia Horticulture Research Station, Alexandria, Egypt. Plant materials of this study were consisted of five genotypes of okra. These genotypes were two varieties and three breeding lines of okra. The varieties were "Sabahia 1" and "Sabahia2", whereas, the lines were originated from a breeding program, started in the summer season of 1995 at the above-mentioned Research Station, Alexandria, Egypt [6].

A randomized complete blocks design was used. Seeds of okra were planted in a single row, 4m long, 0.7 m wide and hills 30 cm apart at the rate of 4 seeds per hill. Each plot contained 5 rows for edible pods yield and 5 rows for seed yield. The plot area was 14 m2 for each. Sowing date was on the first of May. Three weeks later, seedlings were thinned and the strongest one being remained in each hill. Other cultural practices were carried out as recommended for the conventional okra planting. Harvesting took place during the period from mid of June up to mid of September.

Recorded Measurements

Fruit yield and its components: Edible pods were picked with all pedicels in the morning every 3 days interval. The following traits were recorded, total number of edible pods/plant and edible pod weight (g). Pod dry matter (%) was measured as the average of three different pickings and expressed as dry weight/fresh weight x 100. Total edible yield (Kg/fed, was calculated based on the plot area) and number of pods/plant were taken for 25 pickings. At the end of harvesting season, plant height (cm) and number of branches/plant were counted at the end of picking.

Seed yield and its components: At the end of harvesting season, plant height (cm) and number of branches/plant were counted. Number of mature pods/plants, Number of seeds/pod, total seed yield (Kg/fed, was calculated based on the plot area), and average weight of 100 seeds were recorded.

Statistical Procedures

1. Simple statistical analysis of a randomized complete blocks design according to Snedecor and Cochran [7] was done to find out the significance of the studied characters and to compare between means by Duncan's multiple range test at 0.05% level of significance.
2. Simple correlation coefficients (r) were calculated for different pairs of the studied characters as shown by Dospekhove [8].
3. Path-coefficient was calculated as described by Dewey and Lu [3]. In the path diagram (Figure 1), the doubled-arrowed lines indicate mutual association as

measured by correlation coefficients rij the single arrowed lines represent direct influence as measured by pathcoefficient Pix, and "h" represents residual factors.

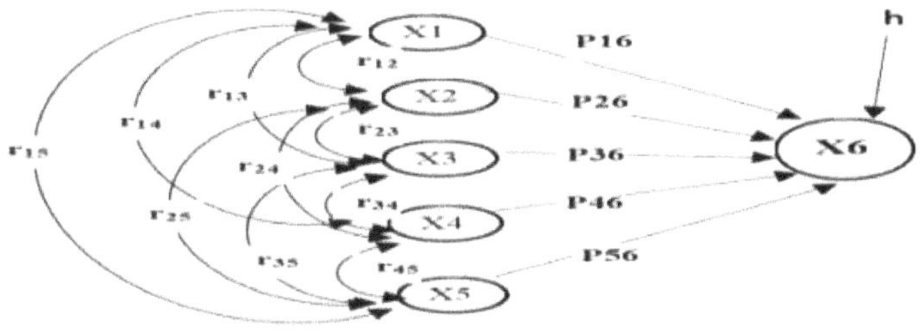

Figure 1: Path diagram with 5 predictor variables "X1" to "X5" and the response variable X6. The variable "h" is the remainder portion or residual. (1-R2) ½

Table 1: Fruits yield and its components of the five okra genotypes, calculated from the combined data over 2013 and 2014 seasons. #Values with the same alphabetical letters, within a comparable group of means, do not significantly differ from one another, using Duncan's multiple range test at 0.05 level of probability

Genotypes	Plant height (cm)	No. of branches / plant	Total No. of edible pods/ plant	Edible pod weight (g)	Dry matter of edible pod (%)	Total edible pod yield (Kg/fed)
Sabahia 1	262.6 a#	10.8 a	81.6 b	3.9 c	16.45 a	5869.2 a
Sabahia 2	259.1 b	7.5 c	59.8 c	4.0 c	16.46 a	4362.7 e
Line 1	240.3 e	9.7 b	59.9 c	4.4 b	16.51 a	4777.0 d
Line 2	241.9 d	9.2 b	83.9 a	3.6 d	16.05 a	5530.1 b
Line 3	252.3 c	6.7 d	49.1 d	5.6 a	14.69 b	5034.4 c

$$P_{16} + r_{12} P_{26} + r_{13} P_{36} + r_{14} P_{46} + r_{15} P_{56} = r_{16} \quad (1)$$

$$r_{21} P_{16} + P_{26} + r_{23} P_{36} + r_{24} P_{46} + r_{25} P_{56} = r_{26} \quad (2)$$

$$r_{31} P_{16} + r_{32} P_{26} + P_{36} + r_{34} P_{46} + r_{35} P_{56} = r_{36} \quad (3)$$

$$r_{41} P_{16} + r_{42} P_{26} + r_{43} P_{36} + P_{46} + r_{45} P_{56} = r_{46} \quad (4)$$

$$r_{51} P_{16} + r_{52} P_{26} + r_{53} P_{36} + r_{54} P_{46} + P_{56} = r_{56} \quad (5)$$

RESULTS AND DISCUSSION

Fruits yield and its components of the five okra genotypesare illustrated in Table 1. Sabahia 1 cultivar had the highest significant mean values for plant height (262.6 cm), no. of branches/plant (10.8), and total edible pod yield (5869.2 kg/fed) comparing with the other genotypes. Meanwhile, the highest significant mean values fortotal number of edible pods/plant (83.9) and edible pod weight (5.6 g) were obtained by Line 2 and Line 3, in order. On the other hand, Line 3 gave the lowest percentage of dry matter in edible pods comparing with the other genotypes. With this respect, Abd-Allah and Mansour [6] established new lines of okra to improve and meet the need of new cultivars of okra for fresh consumption. They selected five lines i.e., L3 had early flowers and the pods were tall and thin; L5 had early flowers and the pods were short and thick; L6 had early flowers with tall and thin; L10 had late flowers with; mild-tall and thick, and L12 had late flower with short and thick. All lines gave tall plants with high values for number of pods and total yield/plant.

Seed yield and its components of the five okra genotypesare shown in Table 2. The highest plants were obtained by the cultivars Sabahia 1 (180.8 cm) and Sabahia 2 (179.1 cm), there were no significance differences between them. Meanwhile, the biggest number of branches/plant were obtained by the cultivar Sabahia 1 (7.1) and Line 2 (6.8), there were no significance differences between them. Line 1 gave the highest significant mean values for number of mature pods/ plant (23.5) and weight of 100 seeds (5.18 g). Line 2 had the highest significant mean values for number of seeds/ pod (86.3) and seed yield (1981.4 kg/fed).These results are according to Abd-Allah [8,9].

Correlation coefficient values among pairs of fruits yield and its components for all okra genotypes are illustrated in Table 3. There were positively correlation between number of branches/plant with both of total edible pod yield and dry matter of edible pod, but it was negatively correlation with edible pod weigh. Total number of edible pods/plant was positively correlated with dry matter of edible pod and highly positively correlated with both of number of branches/plant and total edible pod yield, but highly negatively correlated with edible pod weight. There was highly negatively correlation between edible pod weight and dry matter of edible pod.These results were according with Abd-allah [9-11] and Mansour and Abd- Allah [6].

Table 2: Seed yield and its components of the five okra genotypes, calculated from the combined data over 2013 and 2014 seasons. #Values with the same alphabetical letters, within a comparable group of means, do not significantly differ from one another, using Duncan's multiple range test at 0.05 level of probability

Genotypes	Plant height (cm)	No. of branches /plant	No. of mature pods/plant	No. of seeds/pod	Weight of 100 seeds (g)	Seed yield (Kg/fed)
Sabahia 1	180.8 a#	7.1 a	18.9 c	83.1 b	4.89 b	1753.7 b
Sabahia 2	179.1 a	5.6 c	18.4 c	74.6 e	4.87 b	1549.9 c
Line 1	133.5 d	6.5 b	23.5 a	81.1 c	5.18 a	1750.4 b
Line 2	154.1 c	6.8 a	22.2 b	86.3 a	4.52 c	1918.4 a
Line 3	168.4 b	4.8 d	15.5 d	76.1 d	4.62 c	1590.6 c

Table 3: Correlation coefficient values (r) for pair of the characters of fruits yield and its components of the five okra genotypes. *,** Significant differences at 5% and 1% levels of probability, respectively

Character	Plant height	No. of branches/ plant	Total No. of edible pods/ plant	Edible pod weight	Dry matter of edible pod
Total edible pod yield	0.100	0.633*	0.752**	-0.271 -	-0.029
Dry matter of edible pod	0.042	0.688*	0.511*	0.830**	
Edible pod weight	-0.002	-0.641*	-0.833**		
Total No. edible pods/plant	0.017	0.758**			
No. of branches/ plant	-0.050				

Table 4: Correlation coefficient values (r) for pair of the characters of seed yield and its components of the five okra genotypes. *,** Significant differences at 5% and 1% levels of probability, respectively

Character	Plant height	No of branches /plant	No. of mature pods/plant	No. of seeds/pod	Weight of 100 seeds
Seed yield	-0.472	0.790**	0.718**	0.983**	-0.190
Weight of 100 seeds	-0.342	0.245	0.405	-0.151	
No. of seeds/pod	-0.369	0.851**	0.658*		
No. mature pods/plant	M.764**	0.711**			
No. of branches/plant	-0.206				

Correlation coefficient values among pairs of seed yield and its components for all okra genotypes are illustrated in Table 4. Seed yield had highly positively correlation with number of branches/plant, number of mature pods/plant, and number of seeds/pod. Meanwhile, number of seeds/pod was positively correlated with number mature pods/ plant and highly positively correlated with number of branches/ plant. Moreover, there were highly positive and negative correlation between number of mature pods/plant with number of branches/plant and plant height, in order. In this regard, Abd-Allah [10] suggested that number of branches/plant, and number of pods/plant showed to be the first concern for improving edible and seed yields of okra.

As genetic correlation is the correlation of breeding value and results mainly from pleiotropy [12]. Dealing with the assumption that correlation between yield and its components simply measures mutual relationships without presumption of causation, so that, in pathcoefficient analysis, the researcher specifics the plausible causes and measures their relative importance as independent variables [3,4]. This could be shown from Table 5, where edible pod yield was the product of five traits, i.e., plant height, number of branches/plant, number of edible pods/plant, edible pod weight, and pod dry weight, each of which contributed directly and indirectly to yield.

The direct effect of number of branches/plant was strongly positive (1.222). So, it may indicate a true relationship, and a direct selection through this trait may be effective in improving yield in okra, as reported by Abd-Allah [10]. As correlation coefficient between total number edible pods/plant and edible pod yieldwere positive (Table 3) and the direct effect of these traits was negative (Table 5), it seemed that indirect effect appeared to be the cause of correlation. The residual effect for pod yield/plot was low value (0.007), which means that the studied traits interfere about 99 % of pod yield and 1% of the attributes are needed to be under investigation. In this respect, Ragheb [13] found that the total number of pods/plant had a strong direct effect on yield followed by plant height, and number of branches seemed to have an indirect effect through number of pods/plant. Abd- Allah [11] reported that number of pods/plant was having a stronger direct effect on yield. Meanwhile, number of pods/plant, number of branches/plant and plant height seemed to have an indirect effect through number of pods/plant.

Direct and indirect effects of yield components on seed yield in okra are tabulated in Table 6. The direct effect of number of seed/pod was positive (0.892. Also, it could be noted that the genotypic correlation between of number of seed/pod and seed yield (0.983) seemed to be close to the estimated value of its direct effect, so, it may indicate a true relationship and direct selection through this trait may be effective for improving seed yield of okra.

Table 5: Direct effects (Diagonal) and indirect effects of some studied traits on fruit yield of the five okra genotypes. R = 0.007

Character	Plant height	No. of branches plant	Total No. of edible /plant	Edible pod weight	Dry matter of edible pod	Total effect
Plant height	0.224	-0.061	-0.001	0.002	-0.063	0.100
No. of branches/ plant	-0.011	1.222	-0.052	0.505	-1.031	0.633
Total No. edible pods/ plant	0.004	0.926	-0.069	0.656	-0.765	0.752
Edible pod weight	0.000	-0.783	0.058	-0.788	1.243	-0.271
Dry matter of edible pod	0.009	0.841	-0.035	0.654	-1.498	-0.029

Table 6: Direct effects (Diagonal) and indirect effects of some studied traits on seed yield of the five okra genotypes. R = 0.033

Character	Plant height	No. of branches/ plant	No. of mature pods/ plant	No. of seeds/pod	Weight of 100 seeds	Total effect
Plant height	-0.004	0.028	-0.214	-0.329	0.047	-0.472
No. of branches/ plant	0.001	-0.135	0.199	0.759	-0.034	0.790
No. mature pods/plant	0.003	-0.096	0.280	0.587	-0.055	0.718
No. of seeds/pod	0.001	-0.115	0.184	0.892	0.021	0.983
Weight of 100 seeds	0.001	-0.033	0.113	-0.135	-0.137	-0.190

As the correlation coefficient between number of branches/plant and number of mature pods/plant with seed yield was positive (0.790 and 0.718, in order) and the direct effect of these factors were either negative or negligible, it seems that indirect effect appeared to be the cause of correlation. Hence, these traits may be used simultaneously with the other characters. Abd-Allah [10] reported that number of pods/plant was having a stronger direct effect on yield. Meanwhile, number of pods/plant, number of branches/ plant and plant height seemed to have an indirect effect through number of pods/plant. It worth mentioning that the residual effect for seed yield/plot was low (0.033). Such result indicated that this character may depend on the most of the studied traits, so, it may not be needed to investigate more attributes affecting pod yield in okra.

REFERENCES

1. Berry SK, Kalara CL, Sehgal RC, Kulkarni SG, Sharma BR (1988) Quality characteristics of seeds of five okra (Abelmoschusesculentus L.) cultivars. J FdSciTechnol 25:303-305.
2. Sharma BR (1993) Genetic improvement of vegetable crops. KallooG, Bergh BO (eds.)Pergamon Press Oxford and New York, pp. 751-769.
3. Dewey DR, Lu NH (1959) A correlation and path-coefficient analysis of components of crested wheat grass seed production. Agron J 51:515-518.
4. Li CC (1975) Path-analysis a primer. Boxwood press. Pacific Grove California, pp: 518.
5. Agriyo OJ, Aken'ova ME, Fatokum CA (1987) Plant characters correlation and path analysis of pod yield in okra (Abelmoschusesculentus). Euph 36: 677-686.
6. Abd-Allah SAM, Mansour SMA (2005) Selection for Yield and Its Components in Okra (AbelmoschusEsculentusL. Moench). J AdvAgric Res (Fac. Agric. Saba Bash), 10(4):409-418.
7. Snedecor GW, Cochran WG (1980) Statistical Methods. (7thedn) Iowa State Univ. Press, Ames, Iowa, USA

8. Dospekhov PA (1984) Field experimentations. Statistical Procedures, Mir bublifhebro Moscow, p. 349.

9. Abd-Allah SAM (2004) Stability and Path Coefficient Analyses for Seed Yield and Its Components of Okra [Abelmoschusesculentus (L.) Moench]. The 4th Scientific Conference of Agricultural Sciences. FacAgrAssiutUniv, Assiut, Egypt, 359-369.

10. Abd-Allah SAM (2010) Stability Analyses for Some New Lines of Okra [Abelmoschusesculentus(L.). Minia 2nd ConfAgric Environ Sci 127-138.

11. Abd-Allah SAM (2000) Studies of some aspects on okra breeding (Abelmoschusesculentus (L.) Moench). PhD Thesis, Faculty of Agric. (Saba Pasha) Alexandria University, Alexandria, Egypt.

12. Falconer DS (1981) Introduction to quantitative genetics (2ndedn) The Longman press. Londo, 340.

13. Ragheb WS (1995) Direct and indirect correlation of pod yield and yield components in okra plants (Abelmoschusesculentus (L.) Moench). CominSci and Dev Res 49: 225-234.

14
CHAPTER

CONVENTIONAL AND MOLECULAR MARKER-ASSISTED SELECTION AND PYRAMIDING OF GENES FOR MULTIPLE DISEASE RESISTANCE IN TOMATO

Peter Hanson[a], Shu-Fen Lu[a], Jaw-Fen Wang[a], Wallace Chen[a], Lawrence Kenyon[a], Chee-Wee Tan[a], Kwee Lian Tee [b], Ya-Ying Wang[a], Yun-Che Hsu[a], Roland Schafleitner [a], Dolores Ledesma[a], Ray-Yu Yang[a]

[a] AVRDC—The World Vegetable Center, Shanhua, Tainan 74199, Taiwan

[b] Universiti Malaysia Sabah, School of Sustainable Agriculture, Sungai Batang Mile 10, 90000 Sandakan, Sabah, Malaysia

ABSTRACT

Tomato (Solanum lycopersicum L.) is widely grown in the tropics but production is subject to high losses from diseases. AVRDC—The World Vegetable Center initiated a program to develop fresh market tomato lines resistant to begomoviruses causing tomato yellow leaf curl disease, Phytophthora infestans causing late blight, Ralstonia solanacearum causing bacterial wilt, Stemphyllium spp. causing gray leaf spot, Fusarium oxysporum f. sp. lycopersici race 2, and Tobacco mosaic virus. This work provides greenhouse, field, molecular marker, and laboratory protocols used in the screening and selection process that were applied to segregating populations during generation advance over three years to develop five multiple disease resistant F_7 fresh market tomato lines. Resistance of the five lines to the abovementioned diseases was confirmed in subsequent evaluations. Average yields of the five lines exceeded 100 t/ha under optimal temperatures in a dry season trial, but yields were reduced in a second trial under higher temperatures and rainfall. Seed of three multiple disease resistant $F_{7:8}$ lines is available from AVRDC (http://

Citation: Peter Hanson, Shu-Fen Lu, Jaw-Fen Wang, Wallace Chen, Lawrence Kenyon, Chee-Wee Tan, Kwee Lian Tee, Ya-Ying Wang, Yun-Che Hsu, Roland Schafleitner, Dolores Ledesma, Ray-Yu Yang, Conventional and molecular marker-assisted selection and pyramiding of genes for multiple disease resistance in tomato, Scientia Horticulturae, Volume 201, 30 March 2016, Pages 346-354, ISSN 0304-4238, http://dx.doi.org/10.1016/j.scienta.2016.02.020.

avrdc.org/seed/improved-lines/); these lines have potential for release as inbred line cultivars, hybrid parental lines, or breeding stock.

INTRODUCTION

Tomato (*Solanum lycopersicum* L.) is a widely grown vegetable throughout the tropics and subtropics and is an important source of vitamins A and C. Production of high value fruit and vegetables such as tomato offer some smallholders the opportunity to change from subsistence to commercial farming and substantially increase their incomes (Weinberger and Lumpkin, 2005 and Fan et al., 2013). However, tomato crops can be infected by disease-causing bacterial, fungal, and viral pathogens that reduce yields, fruit quality, shelf-life, and nutritional content. In extreme cases, these diseases force farmers to abandon tomato production altogether. In the absence of resistant cultivars, farmers often depend on pesticides to control diseases. High reliance on pesticides poses health hazards to farmers and their families, the environment, and consumers; intensive pesticide use also can substantially increase production costs, which increase farmer financial risks and pass the accrued higher costs to consumers (Wilson and Tisdell, 2001). Resistant cultivars are among the cheapest, simplest, and most environmentally safe ways to manage disease.

Many diseases affect tomato in the tropics and subtropics, but three of the most important in terms of widespread incidence and potential to cause high yield losses include tomato yellow leaf curl disease caused by whitefly-vectored begomoviruses (Hanssen et al., 2010 and Navas-Castillo et al., 2011), bacterial wilt caused by *Ralstonia solanacearum* (Hayward, 1991 and Mansfield et al., 2012), and late blight caused by*Phytophthora infestans* (Mont.) De Bary (Fry, 2008 and Nowicki et al., 2012). The pathogens causing these diseases are genetically diverse with vast potential to generate new forms (Hayward, 1991 and Fry, 2008). Most disease resistance in commercial tomato cultivars is conditioned by single genes, each conferring resistance to a specific pathogen or pathogen race, strain, or phylotype (Yang and Francis, 2007, Scott and Gardner, 2007 and Scott, 2007). Six tomato yellow leaf curl disease resistance genes (*Ty-1/Ty-3, Ty-2, Ty-4, Ty-5, Ty-6*) are available in cultivated tomato (Ji et al., 2007a, Ji et al., 2007b, Ji et al., 2007c, Verlaan et al., 2013 and Hutton and Scott, 2015). Five late blight resistance genes were introgressed from *S. pimpinellifolium* into cultivated tomato (Nowicki et al., 2012) and *Ph-2* and *Ph-3* have been used in commercial cultivars (Zhang et al., 2014). Two major bacterial wilt resistance quantitative trait loci (QTLs), *Bwr-12* and *Bwr-6*, were identified in tomato cultivar 'Hawaii 7996' (H7996) (Thoquet et al., 1996, Carmeille et al., 2006 and Wang et al., 2013) and *Bwr-12* is important for resistance to Phylotype 1 (Asia) bacterial wilt strains (Wang et al., 2013). Three race-specific genes (*I, I–2, I–3*) condition resistance to the fusarium wilt pathogen (*Fusarium oxysporum* f. sp. *lycopersici*) (Scott and Gardner, 2007). The incompletely dominant gene *Sm* offers resistance to four species of the gray leaf spot pathogen (*Stemphylliumspp.*) (Scott and Gardner, 2007). Several genes condition resistance to *Tobacco mosaic virus* (TMV) in tomato and the $Tm2^2$ allele conditions resistance to multiple strains (Scott, 2007).

Effective selection for disease resistance in segregating populations requires accurate, cost-effective screening methods that permit rapid testing of thousands of plants. Common disease screening techniques include field testing under natural disease pressure, and greenhouse/growth room screening procedures in which plants are inoculated with specific pathogen strains. Field screening is appropriate when the breeding is conducted in the region where the cultivars will be released and high disease pressure can be expected. Greenhouse seedling inoculation can assess disease reactions quickly, reduce some sources of environmental variation by use of characterized pathogen strains and defined inoculum concentrations, and avoid confounding effects from other pests or diseases. Many disease resistance genes have been mapped in tomato, and molecular markers linked to these genes are available for marker-assisted selection (MAS). The choice of screening method depends upon effectiveness, availability, cost, and convenience. Disease resistance alone is insufficient to ensure farmer adoption; commercial cultivars also must possess high yield potential, early maturity, and other horticultural traits, as well as fruit quality and nutrient content. Consequently, disease resistance breeding must be conducted with selection for important horticultural and fruit characters.

Breeding inbred lines with resistance to multiple diseases is a worthy but often difficult goal. Selecting screening protocols and the sequence of trait screening, and managing segregating populations to achieve the desired outcome, can be challenging. This paper describes a three-year selection process and the sequence of field, lab, greenhouse and molecular marker protocols applied by AVRDC-The World Vegetable Center (AVRDC) to a segregating population, which led to the development of fresh market tomato lines resistant to late blight, tomato yellow leaf curl disease, bacterial wilt, fusarium wilt, gray leaf spot, and *Tobacco mosaic virus* (TMV).

MATERIALS AND METHODS

Parents and Cross

A three-parent cross, [(CLN2777G × G2-6-20-15B) × LBR-11], coded CLN3241 was created at AVRDC in 2006–2007 to develop tropically adapted, multiple disease resistant lines. CLN2777G is homozygous for resistance genes *Bwr-12* (bacterial wilt), *Ty-2*(tomato yellow leaf curl disease), and $Tm2^2$ (TMV). G2-6-20-15B is homozygous for *Ty-3*(tomato yellow leaf curl disease resistance). LBR-11 is an AVRDC F_6 selection from North Carolina State University F_2 population NC3220x-20 and is homozygous for resistance genes *Ph-2* and *Ph-3*, *I2* (resistance to race 2 of the fusarium wilt pathogen), and *Sm* (resistance to the gray leaf spot pathogen).

Line Development

Segregating populations were managed by pedigree selection appropriate for self-pollinating crops (Fehr, 1987). Sixteen protocols to assess disease resistance, horticultural and fruit traits (Table 1 and Table 2) were used to screen populations and lines. Selection and generation advance began in 2008 with a segregating triple-cross F_1population (Table 3) and continued until completion of F_7 lines in June 2012. Selection was based on individual plant performance in the F_1 and F_2 generations and single plot progeny rows (30 plants per plot) with individual plant selection practiced within rows from the F_3 to F_7 generations. Two tomato crops were produced in southern Taiwan during the dry season (October–February and March–June) allowing two cycles of generation advance per year. Seed was sown in trays and seedlings were maintained in a plastic house for about 30 days before field transplanting.

Before transplanting, seedlings were screened for resistance to one or more diseases, either by MAS or by greenhouse inoculation. Resistant plants were transplanted to the field for evaluation of fruit and horticultural traits. Intensive selection for tomato yellow leaf curl disease and late blight resistance was practiced during the F_1–F_4 generations. Tomato yellow leaf curl disease in southern Taiwan is caused by *Tomato yellow leaf curl Thailand virus*(TYLCTHV) and *Tomato leaf curl Taiwan virus* (ToLCTWV) (Tsai et al., 2011) with the highest pressure occurring from March to June and coinciding with high whitefly populations. Plants with tomato yellow leaf curl disease resistance genes identified by MAS were assessed in the field for tomato yellow leaf curl disease severity. Late blight screening relied on seedling inoculation with selected pathogen isolates, and also by MAS after gene markers for *Ph-2* and *Ph-3* became available in 2009. Greenhouse seedlings were screened for bacterial wilt resistance using drench inoculation, and by MAS after markers for *Bwr-12* became available in 2011.

Fusarium wilt, TMV, and*Stemplyllium* screenings were performed with seedlings of $F_{7:8}$ lines. Selection for plant vigor, vine cover (extent to which foliage cover protects the fruit), early maturity, visual fruit load, fruit size and shape, fruit size uniformity within the fruit cluster, fruit firmness, fruit color development, and absence of fruit defects such as cracking was carried out during generation advance.

Table 1: Greenhouse, field, molecular marker, and laboratory protocols for selection of tomato disease resistances, horticultural traits, and fruit quality and nutrient contents, AVRDC Taiwan

Trait	Protocol	Protocol no.	Reference
Horticultural	Visual assessment of vine cover, plant habit, fruit set, fruit shape, earliness	1	Hanson et al. (2012)
*Tobacco mosaic*virus resistance	Selection for green seedling hypocotyl conditioned by the *ah* gene (anthocyaninless of Hoffman) linked to*Tm2*^{2}	2	Robinson et al. (1970)
	NCTm-019CAPS marker for *Tm2*^{2}	3	See Table 2
	P1-16 SCAR marker for *Ty-2*	4	See Table 2
	P6-25 SCAR marker for *Ty-3*	5	
Tomato yellow leaf curl disease resistance	Seedling exposure to viruliferous whiteflies	6	Hanson et al. (2012)
	TG328 marker for *Ph-3*	7	See Table 2
	dTG422/Hinfl CAPS marker for *Ph-2*	8	
Late blight resistance	*P. infestans* seedling screening Isolate Pi237 (T1,3) to select for *Ph-2*	9	Chen et al. (2008)
	Isolate Pi39A (T1,2) to select for *Ph-3*	10	
	Seedling drench inoculation	11	Wang et al. (2013)
Bacterial wilt resistance	SLM12-2,SLM12-10 markers for *Bwr-12*	12	See Table 2
	Molecular marker for *I2* gene	13	See Table 2
Fusarium wilt resistance	Fusarium wilt seedling drench inoculation	14	Sheu and Wang (2006)
Gray leaf spot resistance	CT55/Ddel CAPS marker for *Sm*	15	See Table 2
Fruit qualities and nutrients	Color, soluble solids, vitamin C, beta-carotene, lycopene	16	Hanson et al. (2004)

SCAR = sequence characterized regions; CAPS = cleaved amplified polymorphic sequence.

Table 2: Targeted resistance genes, linked molecular markers, and marker sequences used in marker-assisted selection

Gene	Marker name	Marker type	Enzyme	Forward primer sequence 5′–3′	Reverse primer sequence 5′–3′	Annealing temperature (°C)	Reference
Ph-2	dTG422	CAPS		[a]			Mutschler lab
Ph-3	TG328	CAPS	BstN1	GGT GAT CTG CTT ATA GAC TTG GG	AAG GTC TAA AGA AGG CTG GTG C	55	Robbins et al. (2010)
	SLM12-2	SCAR		ATCTCATTCAACGCA-CACCA	AACGGTGGAAAC-TATTGAAAGG	55	Ho et al. (2013)
Bwr-12	SLM12-10	SCAR		ACCGCCCTAGCCATA-AAGAC	TGCGTC-GAAAATAGTTGCAT	55	
Ty-2	P1-16	SCAR		CACACATATCCTC-TATCCTATTAGCTG	CGGAGCTGAATTG-TATAAACACG	55	Yang et al. (2014)
Ty-3	P6-25	SCAR		GGT AGT GGA AAT GAT GCT GCT C	GCT CTG CCT ATT GTC CCA TAT ATA ACC	50	Ji et al. (2007a)
I2	I2OH	SCAR		TGGAGAGTTCCCTA-CACTTGAG	TTCTCTTCAAGG-TAGTTGGCAG	55	Popoola et al. (2014)
Sm	CT55	CAPS	DdeI	CATCTGGTGAGGCGGT-GAAGTA	TCCGCCCAAA-CAAAACAGTAATA	55	Ji and Scott (2009)
Tm2[2]	NCTm-019	CAPS	HaeIII	AATTTGGGCATACT-GACATC	GTTGCACACATTG-GTTGTAG	55	Panthee et al. (2013)

aPrimer sequences for *Ph-2* are available upon request from Dr. Martha Mutschler, College of Agriculture and Life Sciences, Cornell University, USA.

Table 3: Targeted traits, screening sequence, and screening methods applied during generation advance in development of multiple disease resistant tomato lines, AVRDC Taiwan, 2008–2013

Generation	No. plants/population	Period	Targeted traits	[a]Protocol no.
F_1	60 plants	October, 2008–February, 2009	*Tobacco mosaic virus* (TMV) resistance gene *Tm2*[2] by morphological marker	2
			Horticultural (vine cover, plant habit, fruit set, fruit shape, earliness)	1
			Tomato yellow leaf curl disease (TYLCD) resistance genes *Ty-2*, *Ty-3* by marker-assisted selection (MAS) and seedling exposure to viruliferous whiteflies	4,5,6
F_2	4 populations (200 plants/population)	March–June, 2009	Horticultural	1
			TYLCD resistance genes *Ty-2*, *Ty-3* by MAS and plant exposure to viruliferous whiteflies in greenhouse and field	4,5,6
F_3	32 lines	October, 2009–February, 2010	Late blight resistance gene *Ph-3* by MAS and seedling inoculation	7,10
			Horticultural	1
F_4	28 lines	July, 2010–February, 2011	Late blight resistance gene *Ph-2* by MAS and seedling inoculation	8,9
			Horticultural	1

Generation	No. plants/population	Period	Targeted traits	[a]Protocol no.
F_5	36 lines	March–June, 2011	Late blight resistance genes *Ph-2*, *Ph-3* by MAS	7,8
			Horticultural	1
			TYLCD resistance by plant exposure to viruliferous whiteflies in greenhouse and field	6
F_6	59 lines	June–July, 2011	Bacterial wilt resistance by drench inoculation screening	11
F_6	16 lines	October, 2011–February, 2012	Horticultural	1
			Bacterial wilt gene *Bwr-12* by MAS	12
F_7	14 lines	March–June, 2012	Horticultural	1
			Late blight resistance genes *Ph-2*, *Ph-3* by seedling inoculation	9,10
			TYLCD resistance by plant exposure to viruliferous whiteflies in greenhouse and field	6
$F_{7:8}$	5 lines	October, 2012–March, 2014	Late blight resistance genes *Ph-2, Ph-3*seedling inoculation	9,10
			Gray leaf spot resistance gene *Sm* by MAS	15
			Fusarium wilt resistance gene *I2* by MAS and seedling inoculation	13,14
			TMV (*Tm2*[2]) by MAS	3
			Horticultural	1
			Fruit quality, nutrients	16

aProtocols given in Table 1 and Table 2.

Disease resistance Evaluations and Confirmation

Late Blight

Details of the protocols for inoculum preparation and inoculation are given in Chen et al. (2008). Briefly, 35-day-old seedlings were spray-inoculated with zoospore/sporangia suspensions of 5×10^4/mL of selected pathogen isolates. Inoculated seedlings were incubated in a growth room at 100% relative humidity and 20 ± 2 °C without light for the first 24 h. Afterwards, growth room conditions were maintained at 60–95% RH, a daily 14-h light period (70 μE m^2 s^{-1}) and 20 ± 2 °C. Each plant was visually scored 10 days after inoculation according to the following scale, where 0 = no symptoms; 1 = 1–5% leaf area affected, small lesions <2 mm and no stem lesions; 2 = 6–15% leaf area affected, necrosis-restricted leaf lesions and no stem lesions; 3 = 16–30% leaf area affected, coalescing leaf lesions or tiny water-soaked stem lesions; 4 = 31–60% leaf area affected, edge-expanding leaf lesions or a few small stem lesions (< 5 mm); 5 = 61-90% leaf area affected, drying leaf lesions or edge-expanding stem lesions; 6 = 91–100% leaf area affected, leaves blighting, extensive stem damage, or death. Resistant checks included WV700 (homozygous for *Ph-2*) and CLN2037B (homozygous for *Ph-3*). Entries and checks were tested for reactions to isolates Pi39Aand Pi237 in separate trials. WV700 is susceptible and resistant, respectively, to Pi39A and Pi237; conversely, CLN2037B is resistant to Pi39A and susceptible to Pi237. Plots included 12 plants and entries were arranged in a randomized complete block design (RCBD) with three replications.

Bacterial Wilt

A detailed description of the bacterial wilt drench method is given in Hai et al. (2008). Inoculations were conducted with virulent *R. solanacearum* strain Pss4 from Taiwan. Pss4 is classified as Phylotype 1 (Asia), race 1, and biovar 3 (Hai et al., 2008). For inoculum preparation, stored cultures were streaked on tetrazolium chloride medium. Several fluidal colonies were transferred to plates containing 523 media for multiplication at 30 °C for 24 h. Bacterial cells were harvested, suspended in water, and adjusted to OD600 = 0.3, (about 10^8 CFU/mL). Four-week-old seedlings grown in 2-in. pots, approximately at the five-leaf stage, were inoculated without wounding the roots by pouring 20 mL of 1×10^8 inoculum on the soil surface at the base of each plant. Checks included L390 (susceptible) and H7996 (homozygous for *Bwr-12* and *Bwr-6*). Plots included 20 plants and were arranged in a RCBD with three replications. Plants in each plot were evaluated for wilting at weekly intervals for four weeks beginning one week after inoculation. Wilted plants died, so plants were scored as healthy or wilted and the percentage of wilted plants per plot was determined after the last evaluation.

Tomato Yellow Leaf Curl Disease

CLN3241-coded $F_{7:8}$ lines and inbred line checks 'Tanya' (susceptible), CLN2498D (*Ty-2*) and CLN3552B (*Ty-3* and *Ty-2*) were screened in separate nethouse trials for resistance to ToLCTWV or *Tomato yellow leaf curl Thailand virus–Taiwan strain*(TYLCTHV-[TW]). Plots of entries and checks comprised 12 plants and were arranged in a RCBD with two replications. Entries in the TYLCTHV-[TW] and ToLCTWV trials were sown on 11 March and 24 March 2015, respectively. On 30 March, seedlings in the TYLCTHV trial were transferred to a nethouse containing plants of susceptible line CL5915-93D4-1-0-3 infected with TYLCTHV-[TW] and abundant viruliferous whiteflies. Similarly, on 10 April the seedlings in the ToLCTWV trial were transferred to a nethouse containing CL5915-93D4-1-0-3 plants infected with ToLCTWV and abundant viruliferous whiteflies. Every seven days after whitefly exposure, entries in both trials were scored for disease severity using a 1–6 severity scale, where: 1 = healthy, no observable symptoms; 2 = very mild with slight yellowing and mosaic on top leaves and no leaf curling; 3 = mild yellowing, mosaic and/or slight leaf curling on youngest leaves, severe symptoms; 4 = moderate yellowing and/or leaf curling on the youngest (top) leaves; 5 = severe yellowing and blistering and/or severe leaf curling plus some leaf size reduction on the youngest leaves of the main stem and/or at least one branch; 6 = very severe yellowing, blistering and/or very severe leaf curling, leaf deformation, leaf size reduction and stunting.

Fusarium Wilt

The CLN3241-coded $F_{7:8}$ lines and check differentials were assessed in separate trials for resistance to race 1 and race 2 of *F. oxysporum* f. sp. *lycopersici* according to Sheu and Wang (2006). Checks included Bonny Best (susceptible), UC 82-L (race 1 resistance conferred by *I* gene) and Florida MH-1 (*I* gene and race 2 resistance conferred by *I-2* gene). AVRDC isolates Fol-11A (race 1) and Fol-34-1 (race 2) were used for inoculation. The inoculum was prepared by blending 7-day-old potato dextrose agar plates with distilled water (125 mL per plate). Two- week-old tomato seedlings were uprooted and root-dipped in the fungal culture slurry for 5 min and then transplanted to trays containing fresh potting mixture. The inoculated plants were kept in the greenhouse with the mean temperature above 25 °C. Plots of entries and checks included 12 plants and plots were arranged in a RCBD with three replications. Each plant was visually scored three weeks after inoculation according to the following scale, where 0 = healthy; 1 = slight vascular discoloration; 2 = severe vascular discoloration usually with stunting; 3 = plant wilted beyond recovery or dead.

Gray Leaf Spot and TMV

Entries were tested by molecular markers (Table 2).

Genotyping

DNA was isolated from fresh young leaves using the method described by Fulton et al. (1995). Primers and assay conditions for the assessment of the genotypes at resistance gene loci are provided in the references listed in Table 2.

Evaluation of Yield and Horticultural Traits

Two field trials (FT) were conducted at AVRDC Taiwan, the first from October 2012 to March 2013 (FT1), and the second from February–June 2014 (FT2). Entries included five CLN3241 $F_{7:8}$ lines and checks CLN2498D, 'Tanya,' and CLN3078C. Entries were replicated twice and plots were arranged in a randomized complete block design. FT1 was sown and transplanted 24 September, 2012 and 28 October, respectively. FT2 was sown on 8 February and transplanted on 18 March, 2014. Plots included a 1.5 m-wide bed with two 4.0-m-long rows per bed (24 plants). Beds were covered by gray plastic mulch and rice straw and plants were staked and pruned. A basal application of 120N–52P–100K–32 Mg kg/ha and an additional 90N–38P–74K–24 Mg kg/ha was applied over four side dressings. Pesticides were used to control insects and furrow irrigation was applied as needed. FT1 plots were harvested on 5 February, 19 February, 3 March and 18 March, 2013 and FT2 plots were harvested 2 June, 5 June and 16 June, 2014.

Fruit Quality and Nutrient Analyses

Detailed descriptions of fruit quality and nutrient analysis protocols are given in Hanson et al. (2004). About 20 fruit sampled from the second harvest of each plot of each trial were provided to the AVRDC Nutrition lab for analysis of soluble solids, color, ascorbic acid, lycopene, and beta-carotene. Color was measured by a colorimeter on three scales represented as a, b and L. Color values of fresh tomato slurry were calculated as a/b. Soluble solids concentration was measured with a digital refractometer and presented as °brix. Carotenoid analyses were performed using high-performance liquid chromatography (HPLC, Waters, Mass.) equipped with a 717 plus autosampler, 600 controller, 2487 detector (read at 436 nm) with a 125 × 4 mm LiChrospher® 100 RP-18e column, 5 μm (Merck, Darmstadt, Germany) under isocratic conditions at ambient temperatures. The mobile phase was acetoniltrile: methanol (75:25, v/v) at a flow rate of 1.5 mL min^{-1}. Commercial β-carotene and lycopene were used as standards. The determination of total ascorbic acid was on the basis of coupling 2,4-dinitrophenylhydrazine (DNPH) with the ketonic groups of dehydroascorbic acid through the oxidation of ascorbic acid by 2,6-dichlorophenolindophenol (DCPIP) to form a yellow–orange color in acidic conditions. Commercial L-(+)-ascorbic acid was used for calibration.

Statistical Analyses

Data (plot means) of disease evaluation experiments were subjected to analysis of variance appropriate for a RCBD using SAS Online Version 9.4 software (SAS Institute, Inc., Cary, N.C.). Bacterial wilt data (percent healthy plants) were transformed by arc sine square root before analysis. Yield and yield components, horticultural traits, fruit qualities and nutrients were subjected to analysis of variance (ANOVA) for each trial and over trials with the SAS General Linear Model (GLM) procedure. For the combined analysis, a mixed effects model was applied where trials, replications, and trial-by-entry were considered random effects and entry as a fixed effect.

RESULTS

Line Development

Screening, selection, and advance from F_1 to F_7 generations were conducted at AVRDC Taiwan over three years using a combination of field evaluation, MAS, and seedling disease screening (Table 1, Table 2, Table 3 and Table 4) to select for disease resistance, yield, horticultural traits, and fruit qualities. The triple cross segregated for six disease resistances (late blight, tomato yellow leaf curl disease, fusarium wilt, gray leaf spot, TMV, bacterial wilt) and two resistance genes each for late blight (*Ph-2*, *Ph-3*) and tomato yellow leaf curl disease (*Ty-2*, *Ty-3*). Generation advance began with controlled screening of seedlings inoculated in the greenhouse/growth room, or with MAS. The three-week pre-transplant period was sufficient to screen and eliminate susceptible plants, and identify resistant plants for field transplanting. The order and frequency of screening depended on trait importance, structure of the triple cross, ease and reliability of the screening method, and availability of molecular markers.

Disease Resistance Evaluations

Late Blight

Checks CLN2037B (*Ph-3*), WV700 (*Ph-2*), and susceptible parents G2-6-20-15B and CLN2777G showed expected reactions to isolates *Pi*-237 and *Pi*39A after seedling inoculation. The five CLN3241-prefixed lines and LBR-11 were resistant (scores $\leq$2.0) to both isolates (Table 4) and subsequent marker analysis confirmed that these entries were homozygous for *Ph-2* and *Ph-3*.

Table 4: Disease resistances of five F_7-derived F_8 lines, parents, and checks determined by seedling inoculations and/or molecular markers

Entry	Entry type	Tomato yellow leaf curl disease (TYLCD)[1]				Late blight (LB)[2]				Fusarium wilt (FW)[3]			Bacterial wilt (BW)[4]	
		Ty-3	*Ty-2*	ToL-CTWV	TYL-CTHV	*Ph-2*	Pi-237	*Ph-3*	Pi-39A	Fol-11A (Race 1)	Fol-34-1 (Race 2)	*I2*	*Bwr-12*	BW%
				Severity score						DSR			SLM12-2 SLM12-10	
		P6-25	P1-16			dTG422	Score	TG328	Score			I2OH		
CLN3241S	$F_{7:8}$ selection	+	–	2.5 b	2.0 c	+	0.9 d	+	1.8 de	0.0 c	0.3 cd	+	+	35 a–c
CLN3241R	$F_{7:8}$ selection	+	–	2.1 b–d	2.0 c	+	0.8 d	+	1.2 e	0.0 c	0.2 c–e	+	+	47 c
CLN3241P	$F_{7:8}$ selection	+	+	1.2 e	2.0 c	+	0.9 d	+	1.3 e	0.0 c	1.9 b	–	+	35 a–c
CLN3241Q	$F_{7:8}$ selection	+	+	1.5 de	1.8 c	+	1.2 cd	+	1.7 de	0.0 c	0.0 e	+	+	7 ab
CLN324 1H-27	$F_{7:8}$ selection	+	+	1.5 de	2.0 c	+	0.8 d	+	1.4 de	0.0 c	0.03 de	+	+	15 a–c
LBR-11	Parental line	–	–	–	–	+	1.5 c	+	2.2 dc	0.0 c	0.1 c–e	+	–	97 d
G2-6-20-15B	Parental line	+	–	–	–	–	3.9 b	–	3.5 ab	0.0 c	0.03 de	+	–	100 d
CLN2777G	Parental line	–	+	–	–	–	5.2 a	–	3.8 a	0.0 c	1.9 b	–	+	40 bc

Entry	Entry type	Tomato yellow leaf curl disease (TYLCD)[1]				Late blight (LB)[2]				Fusarium wilt (FW)[3]			Bacterial wilt (BW)[4]	
		Ty-3	*Ty-2*	ToL-CTWV	TYL-CTHV	*Ph-2*	Pi-237	*Ph-3*	Pi-39A	Fol-11A (Race 1)	Fol-34-1 (Race 2)	*I2*	*Bwr-12*	BW%
				Severity score						DSR			SLM12-2	
		P6-25	P1-16			dTG422	Score	TG328	Score			I2OH	SLM12-10	
Tanya	TYLCD check (susceptible)	–	9	5.5 a	5.3 a									
CLN2498D	TYLCD check (*Ty-2*gene)	–	+	2.2 bc	3.9 b									
CLN3552B	TYLCD check (*Ty-3*and Ty-2 genes	+	+	2.1 b–d	2.1 c									
H7996	BW check (*Bwr-12*gene)												+	12:00 AM
L390	BW check (susceptible)												–	97 d
WV700	LB check (*Ph-2* gene)					+	1.0 cd	–	2.7 bc					
CLN2037B	LB check (*Ph-3* gene)					–	4.9 a	+	1.5 de	2.1 b	0.1 c–e			
Bonny Best	FW check (susceptible)									3.0 a	2.9 a			

Entry	Entry type	Tomato yellow leaf curl disease (TYLCD)[1]				Late blight (LB)[2]				Fusarium wilt (FW)[3]			Bacterial wilt (BW)[4]	
		Ty-3	*Ty-2*	ToL-CTWV	TYL-CTHV	*Ph-2*	Pi-237	*Ph-3*	Pi-39A	Fol-11A (Race 1)	Fol-34-1 (Race 2)	*I2*	*Bwr-12*	BW%
				Severity score						DSR				
		P6-25	P1-16		______	dTG422	Score	TG328	Score		______	I2OH	SLM12-2 SLM12-10	
UC82-L	FW check (*I*gene)									0.1 c	2.0 b			
MH-1	FW check (*I2* gene)									0.0 c	0.0 e		—	—
Significance of the Entry mean square				**	**		**		**	**	**			**

1*Ty-3* and *Ty-2* genes condition TYLCD resistance and presence (+) or absence (–) was determined by molecular markers P6-25 (*Ty-3*) and P1-16 (*Ty-2*). Marker primer sequences are given in Table 2. ToLCTWV (*Tomato leaf curl Taiwan virus*) and TYLCTHV (*Tomato yellow leaf curl Thailand virus*) are begomoviruses that cause TYLCD. Severity scores were determined six weeks after seedling exposure to viruliferous whiteflies and ranged from 1 (healthy, no symptoms) to 6 (severe stunting, leaf curling, yellowing).

2*Ph-2* and *Ph-3* confer resistant and susceptible reactions, respectively, to pathogen isolate Pi-237; conversely, *Ph-2* and *Ph-3* produce susceptible and resistant reactions, respectively, to isolate Pi39A. Presence (+) or absence (–) of *Ph-2* and *Ph-3* was determined by markers dTG422 and TG328, respectively. Seedlings were inoculated in an AVRDC growth room and disease scores ranged from 0 (no symptoms) to 6 (91–100% of leaf area affected, extensive stem damage, and/or plant dead).

3*I2* resistance gene confers resistance to race 2 of the FW pathogen and its presence (+) or absence (–) was determined by marker I2OH.

Seedlings were screened for FW races 1 and 2 by root-dip inoculation. DSR = disease severity rating where values ranged from 0 (healthy) to 3 (plant wilted beyond recovery or dead).

4*Bwr-12* gene conditions resistance to the BW pathogen and its presence (+) or absence (−) was determined by markers SLM12-2 and SLM12-10. Marker primer sequences are given in Table 2. BW%: is percent wilted plants after drench inoculation with *Ralstonia solanacearum* isolate Pss4. Data were transformed by arc sine square root prior to analysis of variance and mean separation. Non-transformed means are shown.

**Significant at $P < 0.001$.

Bacterial Wilt

Check lines H7996 (resistant) and L390 (susceptible) showed expected reactions after drench inoculation (Table 4). The five F_7 test lines (CLN3241 prefixes) and parent CLN2777G tested positive by marker analysis for *Bwr*-12; % wilted plant means of the five resistant lines ranged from 15 to 35 and all were significantly greater than susceptible parents G2-6-20-15B and LBR-11 with wilted plant means ≥97%.

Tomato Yellow leaf Curl Disease

Susceptible check 'Tanya' developed severe symptoms (severity score >5.0) after infection with TYLCTHV-[TW] or ToLCTWV (Table 4) while CLN2498D developed mild (severity score <3.0) and moderate (severity score ~4.0) symptoms after ToLCTWV and TYLCTHV-[TW] infection, respectively. The five CLN3241 lines (homozygous for both*Ty-3* and *Ty-2* or only *Ty-3*) developed mild (severity score <3.0) symptoms to both begomoviruses.

Fusarium Wilt

The CLN3241-prefixed lines and parents were highly resistant to race 1 based on seedling inoculation (Table 4). The gene *Immunity* (*I*) is common in commercial tomato cultivars and probably all entries were homozygous for *I*, but AVRDC lacks a marker for *I*and its presence could not be confirmed. The five test lines and parents G2-6-20-15B and LBR-11 were resistant to race 2 and also tested positive for the *I-2* gene by marker I2OH.

Gray Leaf Spot

Analysis with marker CT55 revealed presence of the *Sm* resistance gene in parents CLN2777G and LBR-11, and in all CLN3241-coded lines.

TMV

The NCTm-019 marker indicated that parent CLN2777G and the five CLN3241-coded lines were homozygous for *Tm2*^{2}. These lines have green hypocotyls conditioned by the recessive allele 'anthocyaninless of Hoffman' (*ah*), which is in coupling phase linkage with $Tm2^2$ (=$Tm2^a$) (Robinson et al., 1970).

Line Evaluation: Yield, Horticultural Traits, Fruit Quality and Nutrient Content

The combined analysis over trials showed significant entry-trial interactions for most variables and entry means for FT1 and FT2 are presented (Table 5). Analysis of variance revealed significant entry mean squares in both trials for yield, average fruit weight, solids, pH, and vitamin C (data not shown). Entry mean squares for average fruit number per plant, days to maturity, solids, color and beta-carotene were significant or highly significant in FT1 or FT2, but not in both. Mean squares for lycopene and acid were non-significant for both trials. Favorable conditions for tomato production during FT1 (mean daily 26.9 °C/15.4 °C maximum/minimum temperatures; total rainfall of 125 mm) resulted in an average fruit yield of 104 t/ha (Table 5). In contrast, higher temperatures (29.2 °C/21.5 °C maximum/minimum temperatures), higher total rainfall (343 mm) and tomato yellow leaf curl disease pressure during FT2 led to an average mean entry yield of 28 t/ha. Average fruit weights were consistent across trials but a 63% reduction in average fruit number per plant contributed to lower yields in FT2. Number of days to maturity (number of days from transplanting to harvest) was 38 days shorter in FT2 than FT1. Late blight infection and bacterial wilt were absent in both trials but high tomato yellow leaf curl disease symptoms developed on susceptible 'Tanya' and CLN2498D during FT2. Values for solids fell within the normal range for fresh market tomato and were consistent between trials. Mean fruit color (a/b = 1.97) in FT1 indicated a moderately deep red internal color favored by most consumers; however, mean entry fruit color in FT2 (a/b = 1.27) fell by 34%, which was due to a 29% reduction in lycopene content between trials. Lycopene synthesis is temperature sensitive and is significantly reduced above 28 °C (Tomes, 1963). Mean beta-carotene contents (provitamin A) were consistent over trials while vitamin C content was 36% higher in FT2 versus FT1. Vitamin C is an antioxidant and higher temperatures and sunlight during FT2 may have induced greater vitamin C synthesis.

Table 5: Evaluation of multiple disease resistant tomato lines for yield, horticultural, and fruit traits in early and late dry season trials, AVRDC Taiwan

Entry code	Fruit weight (g)		Fruit no. per plant		Maturity days after trans-planting		Yield (t/ha)		Solids (°Brix)		Color[1] a/b		Vitamin C mg/100 g fresh weight		Beta-carotene		Lycopene	
	FT1	FT2	FT1	FT2	FT1	FT2	FT1	FT2	FT1	FT2	FT1	FT2	FT1	FT2	FT1	FT2	FT1	FT2
CLN3241S	131 a	136 a	28	8 d	116 bc	78	116 a	25 bc	4.2	4.7 a	1.78 c	1.27	11.7 bc	19.2 b	0.26	0.33 b	5.93	3.62
CLN3241R	133 a	133 a	31	9 cd	116 bc	78	110 a	25 bc	4.3	4.8 a	1.77 c	1.51	10.4 c	19.6 b	0.31	0.34 b	5.53	4.99
CLN3241P	76 b	84 b	39	19 ab	117 ab	78	111 a	34 a–c	4	3.8 b	1.89 bc	1.15	19.4 ab	20.2 b	0.29	0.28 bc	5.6	3.12
CLN3241Q	75 b	86 b	46	19 ab	117 ab	78	107 a	42 a	3.8	3.6 b	1.88 bc	1.01	16.3 bc	18.7 b	0.26	0.45 a	5.07	2.77
CLN3241H-27	76 b	90 b	43	20 a	116 bc	78	121 a	40 a	3.8	3.8 b	1.84 c	0.97	16.0 bc	20.6 b	0.23	0.30 bc	5.53	3.23
Tanya	69 b	59 c	40	11 cd	118 a	79	65 b	11 e	3.8	4.2 ab	2.05 ab	1.69	12.6 bc	18.5 b	0.48	0.22 c	5.6	6.04
CLN2498D	70 b	49 c	41	12 b–d	115 c	80	101 a	16 de	3.7	3.5 b	2.14 a	0.97	16.8 bc	23.3 b	0.41	0.29 bc	7.34	2.58
CLN3078C	82 b	88 b	35	16 a–c	118 a	79	101 a	34 ab	4.2	4.8 a	2.08 a	1.64	27.4 a	37.7 a	0.4	0.34 b	5.06	5.69
Mean	89	90	38	14	116	78	104	28	3.95	4.1	1.93	1.27	16.3	22.2	0.33	0.32	5.71	4
Entry mean square	**	**	ns	*	*	ns	*	**	ns	*	**	ns	*	**	ns	*	ns	ns

FT = field trial. FT1 (dry season) and FT2 (early rainy season) were conducted October 2012–March 2013 and February–June 2014, respectively.

Means within columns followed by the same letter are not significantly different by least significant difference ($P < 0.05$).

1Values for *a* and *b* were measured with a chromometer using a red standard surface. Immature green tomatoes have *a*/*b* ratio less than 0. The *a*/*b* ratio increases to zero and above as the fruits ripen toward a dark red. Values > 2.0 have superior color.

*Significant at $P < 0.05$.

**Significant at $P < 0.01$.

CLN3241S, CLN3241R, CLN3241P, CLN3241Q, and CLN3241-H27 have semi-determinate plant habits producing fair to good foliage cover; moderate pruning is recommended depending on plant spacing. Each line produced deep-globe shaped, moderately firm fruit with uniform (*u* gene) shoulders. Fruit pedicels were jointed. Fruit of CLN3241S and CLN3241R had 5–7 locules, while CLN3241H-27 and CLN3241P had 2–3 locules. All lines tended to develop rain check (small cracks in the cuticle) under high rainfall and relative humidity.

DISCUSSION

This study describes the integrated application of molecular markers and field and greenhouse screening methods including appropriate checks to develop multiple disease resistant tomato lines. This achievement was possible through multidisciplinary collaboration between breeders, plant pathologists, and molecular breeders. The multiple disease resistant tomato lines could be valuable in mid-altitude tropical production areas such as the East African highlands, parts of south India, or Central America, where late blight, tomato yellow leaf curl disease, and bacterial wilt commonly occur sequentially or simultaneously. Yields of the five lines were high under optimal temperatures but much reduced under high temperatures and rainfall, and thus are not recommended for hot rainy season production in the lowland tropics. Seed of three of the five lines is available from AVRDC (http://avrdc.org/seed/improved-lines/); these lines have potential for release as inbred line cultivars, hybrid parental lines, or as breeding stock.

Effective and inexpensive screening methods enable efficient selection of plants carrying desired combinations of resistance genes. Different screening methods were employed during generation advance including field evaluation under natural disease pressure, seedling inoculation with selected pathogen isolates, one morphological marker, and DNA markers linked to disease resistance genes. AVRDC tomato disease resistance screening has been based mainly on seedling inoculation, but MAS has become increasingly important. Molecular markers enabled discernment of plants with both *Ty-3*and *Ty-2* genes versus plants with *Ty-3* alone—a distinction not possible based on symptoms alone after exposure to the leaf curl begomoviruses prevalent in Taiwan. Field screening under high tomato yellow leaf curl disease pressure was important to confirm that lines with *Ty* resistance genes expressed high levels of resistance. Late blight screening in an early generation of this population was performed by sequential screening with selected pathogen isolates targeting *Ph-2* and *Ph-3* resistance genes. Molecular markers for *Ph-2* and *Ph-3* were used for screening after they were made available to AVRDC by Martha Mutschler of Cornell University, and close linkage between the markers and resistance genes was confirmed. The *Ph-2* and *Ph-3* markers were advantageous because individual plants could be assayed for both *Ph-2* and *Ph-3*,and homozygotes versus heterozygotes could be distinguished. Expression of bacterial wilt resistance is affected by pathogen strain and environmental factors such as temperature, moisture, and nematode infection (Hayward, 1991). In the past, selection for bacterial wilt resistance at AVRDC was delayed until the F_3 or later generations, when the lines were screened multiple times during generation advance to confirm resistance. Availability of markers for *Bwr-12* enabled identification of lines carrying this resistance gene.

Broad and durable resistance to a wide range of pathogen races/strains is a desired but often elusive objective of resistance breeding. Durability of resistance is usually known only in hindsight; it is affected by multiple factors such as the extent and time to which a particular resistance gene has been commonly deployed in cultivars, the frequency in which production environments favor the pathogen and disease (Kang et al., 2005), and the factors affecting pathogen populations such as mutation rates, gene flow, reproduction and mating systems, spore persistence, and selection pressure (McDonald and Linde, 2002). The partially dominant gene *Sm* that confers resistance to four*Stemphyllium* species has been incorporated in many tomato cultivars grown throughout the world without reports of resistance breakdown for more than 60 years (Parlevliet, 2002). TMV resistance conferred by the single dominant gene $Tm2^2$ (or $Tm2^a$) also has remained durable (Harrison, 2002, Kang et al., 2005 and Scott, 2007) despite many years of widespread use in tomato cultivars. $Tm2^2$ inhibits viral cell-to-cell movement and two mutations in the pathogen are required to break resistance (Harrison, 2002).

Durable resistance to the late blight pathogen and begomoviruses poses major challenges to breeders and pathologists. Both pathogens are highly diverse, with the capacity to create new forms through recombination and migration. Consequently, a gene pyramiding strategy to combine multiple and complementary resistance genes has been suggested to improve chances of conferring durable resistance to these diseases (Ji et al., 2007b, Vidavski et al., 2008 and Nowicki et al., 2012). All CLN3241-prefixed lines are homozygous for the incompletely dominant gene *Ty-3* and three lines were homozygous for *Ty-3* and *Ty-2*. *Ty-3* exhibited high levels of resistance to *Tomato yellow leaf curl Thailand virus* (TYLCTHV) and *Tomato leaf curl Taiwan virus* (ToLCTWV), multiple bipartite begomoviruses in Guatemala (Garcia et al., 2008), and to monopartite and bipartite begomoviruses in India (Prasanna et al., 2014). AVRDC gives high priority to incorporation of *Ty-3* into its new breeding lines. Our results indicate that *Ty-3* reduced tomato yellow leaf curl disease symptom severity, but did not prevent virus infection. Blocking virus infection by pyramiding *Ty-3* with additional *Ty* genes is desirable to reduce begomovirus inoculum and possible co-infection by multiple viruses and potential recombination. The *Ty-3* + *Ty-2* combination did not eliminate virus infection in this study and pyramiding new resistance gene combinations such as *Ty-3*, *Ty-5* and *Ty-2* should be explored.

P. infestans is notorious for its capacity to quickly overcome resistant cultivars (Fry, 2008). Pathogen diversity and disease pressure will increase with migration and introduction of new clonal lineages and the A2 mating type (Fry, 2008). The CLN3241-prefixed lines are homozygous for the dominant gene *Ph*-3 and the partially dominant gene *Ph-2*. Both genes are race-specific (Chen et al., 2008), but *Ph-2* to a lesser extent than *Ph-3* (Nowicki et al., 2012). Their resistances complement each other in pathogen race protection (Chen et al., 2008). Resistance and durability of cultivars with *Ph-2* and*Ph-3* will depend on the prevalent pathogen clonal lineages in a particular region and whether both A1 and A2 mating types are present. Tomato cultivars with *Ph-3* alone have been effective in Tanzania (Ojiewo et al., 2010) and probably other East African countries where *P. infestans* isolates are derived from the US-1 clonal lineage (Pule et al., 2013). Cultivars with *Ph-2* and *Ph*-3 resistance generally have held up well in the USA (Mutschler, personal communication) but not in Taiwan after displacement of US-1 by isolates of the US-11 clonal lineage, beginning around 2004 (Chen et al., 2008). Tomato breeders continue to search for new late blight resistance genes in wild tomato species*S. pimpinellifolium* (Merk et al., 2012) and *S. habrochaites* (Brouwer and Clair, 2004 and Li et al., 2011), but introgression of new alleles from wild species into commercial cultivars requires many years. Farmers should be encouraged to adopt sanitation, fungicides, and other control tactics to help prolong the durability of late blight resistance.

Linkage of resistance genes can facilitate or hinder gene pyramiding, depending on whether resistances are linked in a coupling or repulsion phase. The *I-2* gene for resistance to race 2 of the fusarium wilt pathogen was mapped to the bottom of chromosome 11 (Hanson et al., 2000 and Ji et al., 2009,) and it is located about 400,000 base pairs below the *Ty-2* locus for begomovirus resistance (Yang et al., 2014). The parental lines of the triple cross used in this study carried these genes in a repulsion phase configuration with genotypes *Ty-2/i2*$^{+}$ (CLN2777G) or *ty-2*$^{+}$*/I2* (G2-6-20-15B, LBR-11). Crossover between the *Ty-2* and *I–2* loci occurred and resulted in one or more recombinants in a coupling phase linkage (*Ty-2/I2*); CLN3241H-27, CLN3241P, and CLN3241Q were homozygous for *Ty-2* and *I2*. MAS has been applied to identify coupling phase resistance gene recombinants, such as was done for *Ph-3* and *Sw-5* (Robbins et al., 2010). Identification of *Ty-2* and *I2* recombinants in this study did not result from a directed search, and the coupling phase linkage was not detected until the $F_{7:8}$ lines were screened for fusarium wilt resistance; the *Ty-2* and *I2* crossover event would have been rare (Yang et al., 2014). The three lines homozygous for *Ty-2* and *I2* were derived from the same F_2 plant, so the crossover occurred in the F_2 or F_1. This is the first report of tomato lines with *Ty-2/I2* in the coupling phase linkage—a significant finding, because both genes will co-segregate in future crosses.

CONCLUSIONS

The multiple disease resistance program described here began in 2007 when selection at AVRDC relied mainly on field and greenhouse disease screening. Molecular markers linked to the targeted resistance genes became available during the three-year line development period and confidence in the markers increased after multiple comparisons between marker and seedling screening results. Effective markers are now available for all disease resistance genes targeted in this study and for many other important genes. AVRDC and many tomato breeding programs routinely apply MAS to large segregating populations, especially the F_2, which increases breeding efficiency through identification of multiple homozygotes and early elimination of susceptible genotypes, and focuses labor-intensive and costly field evaluation on high potential lines. Nevertheless, biological screening to confirm disease resistance in advanced inbred lines will still be a critical step in the development of multiple disease resistant tomato.

ACKNOWLEDGEMENTS

The authors thank Rose Chan for formatting assistance and Maureen Mecozzi for editing the manuscript. Financial support from the Federal Ministry for Economic Cooperation and Development, Germany, project number 11.7860.7-001.00 and 03.7860.4-001.00 is gratefully acknowledged. Core funding to support AVRDC activities worldwide is provided by the Republic of China (ROC), UK Department for International Development (UK/DFID), United States Agency for International Development (USAID), Germany, Thailand, Philippines, Korea, and Japan.

REFERENCES

1. Brouwer, D.J., St. Clair, D.A., 2004. Fine mapping of three quantitative trait loci for late blight resistance in tomato using near isogenic lines (NILS) and sub-NILS. Theor. Appl. Genet. 108, 628–638.
2. Carmeille, C., Caranta, C., Dintinger, J., Prior, P., Luisetti, J., Besse, P., 2006.
3. Identification of QTLs for Ralstonia solanacearum race 3-phylotype II resistance in tomato. Theor. Appl. Genet. 113, 110–121.
4. Chen, C.-H., Sheu, Z.-M., Wang, T.-C., 2008. Host specificity and tomato-related race composition of Phytophthora infestans isolates in Taiwan during 2004 and 2005. Plant Dis. 92, 751–755.
5. Fan, S., Brzeska, J., Keyzer, M., Halsema, A., 2013. From Subsistence to Profit: Transforming Smallholder Farms. Food Policy Report. International Food Policy Research Institute, Washington D.C.
6. Fehr, W., 1987. Principles of Cultivar Development, Vol. 1. Macmillan Publishing Co. NY, USA, 332–338.
7. Fry, W.E., 2008. Phytophthora infestans: the plant (and R gene) destroyer. Mol. Plant Pathol. 9, 385–402.
8. Fulton, T.M., Chunwongse, J., Tanksley, S.D., 1995. Microprep protocol for extraction of DNA from tomato and other herbaceous plants. Plant Mol. Biol. Rep. 13, 207–209.
9. Garcia, B.E., Mejía, L., Melgar, S., Teni, R., Sánchez-Pérez, A., Barillas, A.C., Montes, L., Keuler, N.S., Salus, M.S., Havey, M.J., Maxwell, D.P., 2008. Effectiveness of the Ty-3 introgression for conferring resistance in F3 families of tomato to bipartite begomoviruses in Guatemala. Rep. Tomato Genet. Coop. 58, 22–28.
10. Hai, T.T.H., Esch, E., Wang, J.-F., 2008. Resistance to Taiwanese race 1 strains ofRalstonia solanacearum in wild tomato germplasm. Eur. J. Plant Pathol. 122,471–479.
11. Hanson, P.M., Bernacchi, D., Green, S., Tansksley, S.D., Muniyappa, V., Padmaja, A.S., Chen, H., Kuo, G., Fang, D., Chen, J., 2000. Mapping a wild tomato introgression associated with tomato yellow leaf curl virus resistance in a cultivated tomato line. J. Am. Soc. Hortic. Sci. 125, 15–20.

12. Hanson, P.M., Yang, R.-Y., Wu, J., Chen, J.-T., Ledesma, D., Tsou, S.C.S., Lee, T.-C., 2004. Variation for antioxidant activity and antioxidants in tomato. J. Am. Soc. Hortic. Sci. 129, 704–711.
13. Hanson, P., Lin, L.-J., Luther, G.C., Tsai, W.-S., Srinivasan, R., Chen, C.-H., Lin, C.-H., Sheu, Z.-M., Lu, S.-F., 2012. International Cooperators' Guide: Procedures for Tomato Variety Field Trials. AVRDC Publication, pp. 10–735.
14. Hanssen, I.M., Lapidot, M., Thomma, B.P.H.J., 2010. Emerging viral diseases of tomato crops. Mol. Plant Microbe Inter. 23, 539–548.
15. Harrison, B.D., 2002. Virus variation in relation to resistance-breaking in plants. Euphytica 124, 181–192.
16. Hayward, A.C., 1991. Biology and epidemiology of bacterial wilt caused by Pseudomonas solanacearum. Ann. Rev. Phytopathol. 29, 65–87.
17. Ho, F.-I., Chung, C.-Y., Wang, J.-F., 2013. Distribution of major QTLs associated with resistance to Ralstonia solanacearum phylotype 1 strain in a global set of resistant tomato accessions. Rep. Tomato Genet. Coop. 63, 22–30.
18. Hutton, S.F., Scott, J.W., 2015. Ty-6, a major begomovirus resistance gene located on chromosome 10. Rep. Tomato Genet. Coop. 64, 14–18.
19. Ji, Y., Scott, J.W., 2009. A CAPS marker linked to the tomato gray leafspot (Stemphyllium sp.) resistance gene Sm. Rep. Tomato Genet. Coop. 59, 29–31.
20. Ji, Y., Schuster, D.J., Scott, J.W., 2007a. Ty-3, a begomovirus resistance locus near the tomato yellow leaf curl virus resistance locus Ty-1 on chromosome 6 of tomato. Mol. Breed. 20, 271–284.
21. Ji, Y., Salus, M., van Betteray, B., Smeets, J., Jensen, K.S., Martin, C.T., Mejia, L., Scott, J.W., Havey, M.J., Maxwell, D.P., 2007b. Co-dominant SCAR markers for detection of the Ty-3 and Ty-3a loci from Solanum chilense at 25 cM of
22. chromosome 6 of tomato. Rep. Tomato Genet Coop. 57, 25–28.
23. Ji, Y., Scott, J.W., Hanson, P., Graham, E., Maxwell, D.P., 2007c. Sources of resistance, inheritance, and location of genetic loci conferring resistance to members of the tomato-infecting begomoviruses. In: Czosnek, H. (Ed.), Tomato Yellow Leaf Curl Virus Disease: Management, Molecular Biology, Breeding for Resistance. Kluwer, Dordrecht, 343–362.
24. Ji, Y., Scott, J.W., Schuster, D.J., 2009. Towards fine mapping of the Tomato yellow leaf curl virus resistance gene Ty-2 on chromosome 11 of tomato. HortScience. 44, 614–618.
25. Kang, B.-C., Yeam, I., Jahn, M.M., 2005. Genetics of plant virus resistance. Ann. Rev. Phytopathol. 43, 581–621.
26. Li, J., Liu, L., Bai, Y., Finkers, R., Wang, F., Du, Y., Yang, Y., Xie, B., Visser, R.G.F., van Heusden, A.W., 2011. Identification and mapping of quantitative resistance to late blight (Phytophthora infestans) in Solanum habrochaites LA1777. Euphytica 179, 427–438.

27. Mansfield, J., Genin, S., Magori, S., Citovsky, V., Sriariyanum, M., Ronald, P., Dow, M., Verdier, V., Beer, S.V., Machado, M.A., Toth, I., Salmond, G., Foster, G.D., 2012.
28. Top 10 plant pathogenic bacteria in molecular plant pathology. Mol. Plant Pathol. 13, 614–629.
29. McDonald, B.A., Linde, C., 2002. The population genetics of plant pathogens and breeding strategies for durable resistance. Euphytica 124, 163–180.
30. Merk, H.L., Ashrafi, H., Foolad, M.R., 2012. Selective genotyping to identify late blight resistance genes in an accession of the tomato wild species Solanum pimpinellifolium. Euphytica 187, 63–75.
31. Navas-Castillo, J., Fiallo-Olivé, E., Sánchez-Campos, S., 2011. Emerging virus diseases transmitted by whiteflies. Ann. Rev. Phytopathol. 49, 219–248.
32. Nowicki, M., Foolad, M.R., Nowakowska, M., Kozik, E.U., 2012. Potato and tomato late blight caused by Phytophthora infestans: an overview of pathology and resistance breeding. Plant Dis. 96, 4–17.
33. Ojiewo, C.O., Swai, I.S., Olouch, M.O., Silué, D., Nono-Womdim, R., Hanson, P., Black, L., Wang, T.-C., 2010. Development and release of late blight-resistant tomato varieties 'Meru' and 'Kiboko'. Int. J. Veg. Sci. 16, 134–147.
34. Panthee, D.R., Brown, A.F., Yousef, G.G., Ibrahim, R., Anderson, C., 2013. Novel molecular marker associated with Tm2a gene conferring resistance to tomato mosaic resistance virus in tomato. Plant Breed. 132, 413–416.
35. Parlevliet, J.E., 2002. Durability of resistance against fungal, bacterial, and viral pathogens; present situation. Euphytica 124, 147–156.
36. Popoola, A.R., Ercolano, M.R., Feriello, F., Kaledzi, P.D., Kwoseh, C., Ganiyu, S.A., Ojo, D.K., Adegbite, D.A., Falana, Y., 2014. CAPS markers TAO1 and TG105 in the identification of I2 resistant gene in Nigerian accessions of tomato, Solanum lycopersicum L. Niger. J. Biotech. 28, 43–51.
37. Prasanna, H.C., Sinha, D.P., Rai, G.K., Krishna, R., Kashyap, S.P., Singh, N.K., Singh, M., Malathi, V.G., 2014. Pyramiding Ty-2 and Ty-3 genes for resistance to monopartitie and bipartite tomato leaf curl viruses in India. Plant Pathol. 64,256–264.
38. Pule, B.B., Meitz, J.C., Thompson, A.H., Linde, C.C., Fry, W.E., Langenhoven, S.D., Meyers, K.L., Kandolo, D.S., van Rij, N.C., McLeod, A., 2013. Phytophthora infestans populations in central, eastern and southern African countries consistof two major clonal lineages. Plant Pathol. 62, 154–165.
39. Robinson, R.W., Provvidenti, R., Schoeder, W.T., 1970. A marker gene for tobacco mosaic virus resistance. Rep. Tomato Genet. Coop. 20, 55–56.
40. Robbins, M.D., Masud, M.A.T., Panthee, D.R., Gardner, R.G., Francis, D.M., Stevens, M.R., 2010. Marker-assisted selection for coupling phase resistance to Tomato spotted wilt virus and Phytophthtora infestans (late blight) in tomato. Hortic. Sci.45, 1424–1428.

41. Scott, J.W., 2007. Breeding for resistance to viral pathogens. In: Razdan, M.K., Mattoo, A.K. (Eds.), Genetic Improvement of Solanaceous Crops Vol. 2: Tomato. Science Publishers, Enfield NH, USA, 457–485.

42. Scott, J.W., Gardner, R.G., 2007. Breeding for resistance to fungal pathogens. In: Razdan, M.K., Mattoo, A.K. (Eds.), Genetic Improvement of Solanaceous Crops Vol. 2: Tomato. Science Publishers, Enfield NH, USA, pp. 421–456.

43. Sheu, Z.M., Wang, T.-C., 2006. First report of race 2 of Fusarium oxysporum f. sp. lycopersici, the causal agent of fusarium wilt on tomato in Taiwan. Plant Dis. 90, 111.

44. Thoquet, P., Olivier, J., Sperisen, C., Rogowsky, P., Laterrot, H., Grimsley, N., 1996. Quantitative trait loci determining resistance to bacterial wilt in tomato cultivar Hawaii7996. Mol. Plant Microbe Inter. 9, 826–836.

45. Tomes, M.L., 1963. Temperature inhibition of carotene synthesis in tomato. Bot. Gaz. 124, 180–185.

46. Tsai, W.S., Shih, S.L., Kenyon, L., Green, S.K., Jan, F.-J., 2011. Temporal distributionand pathogenicity of the predominant tomato-infecting begomoviruses in Taiwan. Plant Pathol. 60, 787–799.

47. Vidavski, F., Czosnek, H., Gazit, S., Levy, D., Lapidot, M., 2008. Pyramiding of genes conferring resistance to Tomato yellow leaf curl virus from different wildtomato species. Plant Breed. 127, 625–631.

48. Verlaan, M.G., Hutton, S.F., Ibrahem, R.M., Kormelink, R., Visser, R.G.F., Scott, J.W., Edwards, J.D., Bai, Y., 2013. The tomato yellow leaf curl virus resistance genes Ty-1 and Ty-3 are allelic and code for DFDGD-Class RNA-dependent RNA polymerases. PLoS Genet. 9, 1–11.

49. Wang, J.-F., Ho, F.-I., Truong, H.T.H., Huang, S.-M., Balatero, C.H., Dittapongpitch, V., Hidayati, N., 2013. Identification of major QTLs associated with stable resistance of tomato cultivar 'Hawaii 7996' to Ralstonia solanacearum.

50. Euphytica 190, 241–253.

51. Weinberger, K., Lumpkin, T.A., 2005. Horticulture for Poverty Alleviation—the Unfunded Revolution Shanhua. Taiwan: Avrdc-the World Vegetable Center, AVRDC Publication No. 05-613, Working Paper No. 15.

52. Wilson, C., Tisdell, C., 2001. Why farmers continue to use pesticides despite environmental, health, and sustainability costs? Ecol. Econ. 39, 449–462.

53. Yang, W., Francis, D.M., 2007. Genetics and breeding for resistance to bacterial diseases in tomato: prospects for marker-assisted selection. In: Razdan, M.K., Mattoo, A.K. (Eds.), Genetic Improvement of Solanaceous Crops Vol. 2: Tomato.

54. Science Publishers, Enfield NH, USA, 379–419.

55. Yang, X., Caro, M., Hutton, S.F., Scott, J.W., Guo, Y., Wan, X., Rashi, M.H., Szina, D., de Jong, H., Visser, R.G.F., Bai, Y., Du, Y., 2014. Fine mapping of the tomato yellow leaf curl virus resistance gene Ty-2 on chromosome 11 of tomato. Mol. Breed.34, 749–760.

56. Zhang, C., Liu, L., Wang, X., Vossen, J., Li, G., Li, T., Zheng, Z., Gao, J., Guo, Y., Visser, R.G.E., Li, J., Bai, Y., Du, Y., 2014. The Ph-3 gene from Solanum pimpinellifolium encodes CC-NBS-LRR protein conferring resistance to Phytophthora infestans. Theor. Appl. Genet. 127, 1353–1364.

15
CHAPTER

GENETIC ALGORITHM: A VERITABLE TOOL FOR SOLVING AGRICULTURAL EXTENSION AGENTS TRAVELLING PROBLEM

Adewumi, I.O[1] , Oluwatoyinbo, F.I[2] ,Ajisegiri, G.O[3] and Akinsete, A.E[4]

[1] Department of Agricultural Engineering, Federal College of Agriculture, I.A.R & T P.M.B 5029 Moor Plantation Ibadan, Oyo-State, Nigeria

[2] Department of Agriculture, Federal College of Agriculture, I.A.R & T P.M.B 5029 Moor Plantation Ibadan, Oyo-State, Nigeria

[3] Department of Industrial and Production Engineering, Faculty of Technology, University of Ibadan, Nigeria

[4] Department of Electrical and Electronics, Faculty of Technology, University of Ibadan, Nigeria

ABSTRACT

Genetic algorithm simulates the logic of Darwinian selection as observed in the biological evolutionary process (Cells' division, DNA, Mutation, etc) to solve problems. They are based on one hand on a heuristic gradient ascension method (selection and crossover) and in another hand on a semi-random exploration method (Mutation). In this research work, application of genetic algorithms was explored for the optimization problem embodied in the transit problem of agricultural extension agents or workers in disseminating new innovation and technological advancement in agriculture. An order representation for the cost matrix for 10 cities and chromosomes was used. The result revealed that genetic algorithm can solve the routing problem of an agricultural extension agents in terms of time minimization in order to search for the shortest route, which will increase number of places that the extension agents can touch at reduced cost of transportation. This will help in achieving the nations' vision 2020 on food security.

Citation: Adewumi, I.O, Oluwatoyinbo, F.I ,Ajisegiri, G.O and Akinsete, A.E, Genetic Algorithm: A Veritable Tool For Solving Agricultural Extension Agents Travelling Problem. ISBN: 978-2902-58-6 Volume3.

INTRODUCTION

Sebusang and Masupe, 2003;Torimuno et al., 2007 as cited by Odoemelam and Onuekwusi, 2013 that the crucial role of Information and Communication Technologies cannot be overemphasized in todays' world globalization, perceived as a major driver of the worlds' economies, is powered by an enhanced use of ICTs on which virtually every sector now relies. The current dispensation of competiveness within the global market space could not have been made possible without the interconnectedness of institutions and people. But many factors have continued to impinge on the increase drive towards the use of ICTs in developing economies particularly in Nigeria. Adewumi, 2009 has stated that the need to solve optimization problems is a dominant theme in the engineering world. A great number of analytic and numerical optimization techniques have been developed, and yet there are still large groups of functions that present significant difficulties for numerical techniques, and moreover, for analytical methods these are not continuous or differentiable everywhere; functions, which are non-convex, multi-modal, and functions which contain noise. As a consequence, there is a continuing search for more robust optimization techniques that are able to overcome such problems. Oliver et al (1987) has described Genetic Algorithm (GA) has one of the relatively new class of stochastic search algorithms. Stochastic algorithms are those that use probability to help their search. As the name implies, GA behaves much like biological genetics. It encodes information into strings just as living organisms encode characteristics into strands of DNA. Davies (1987) has revealed in Adewumi (2009) stated that a string in a GA is analogous to a chromosome in biology. A population of strings competes and those strings that are the fittest procreate, the rest eventually die off childless. As with biological parents, two strings combine and contribute part of their characteristics to create their offspring, the new individual. This new string joins the population and fight to produce the next generation. If both parents contribute good building blocks (short section of the string) to the offspring, it will be fit and will procreate in its turn. If the building blocks are poor, then the offspring will die off without generating offspring. A second-but-important process occurs in Gas: sometimes vary rarely. A mutation occurs and the offspring will incorporate a new building block that came from neither parent. The above cycle of death and birth repeats until an acceptable solution to the problem is found (Pohlheim, 2003). Goldberg (1989) as reported by Pransana et.al (1991) defines genetic algorithm (GA) as search algorithm based on mechanics of natural selection and natural genetics. Frick (1998) gave a similar definition stating that GA is a software procedure modeled after genetics and evolution. Genetic algorithm is an adaptive heuristic search algorithm premised on the evolutionary ideas of natural selection and genetics. The basic concept of GA is designed to stimulate processes in natural system necessary for evolution, specifically those that follow the principle laid down by Charles Darwin of survival of the fittest. As such, it represents an intelligent exploitation of a random search within a defined search space to solve a problem. First pioneered by John Holland in the 1960s, genetic algorithm has been widely studied, experimented and applied in many fields of engineering world. Not only does it provide alternative methods to solving problem, it consistently outperforms other traditional methods in most of the problem links. Many of the real world problems involved finding optimal parameters, which might prove difficult for traditional methods, but ideal for GA (De

Jong, 1993); GA exploits the idea of the survival of the fittest and an interbreeding population t create a novel and innovative search strategy. A population of strings, representing solutions to a specified problem is maintained by the GA.

REPRESENTATION OF THE PROBLEM

Domain Genetic algorithm does not deal with the data contained in the problem (or its possible solutions directly, but uses a representation of these data. This representation is mostly an encoding of the original numerical (decimal) values into a binary string of '0' and '1'. In theory, the encoding can be done over any form of finite alphabets, but the binary representation is the most efficient as it is the most basic form of representation. Very often, the problem or its solution contains data not in numerical form. A behavior-oriented model for instance, will contain possible behavior rules the system is looking to optimize. A trading model for financial markets (such as Adaptive Portfolio Trading (APT) system) contains the behavior rules of 'buy' and 'sell' that the GA processes should learn to select once certain conditions are met. A GA-ready model will typically represent these rules through enumerative integer values (as simply as or data types). When the GA engine returns the selected integer value, the application object will execute the appropriate rule normally using some kind of a or nested statement.

Biological Metaphors for Genetic Algorithm

Genetics Within most cells in the human body (and in most living organisms) are rod-like structures called Chromosomes. These chromosomes dictate various hereditary aspects of the individual. Within the chromosomes are individual genes: a gene encodes a specific feature of the individual, for example, a person's eye color is dictated by a specific gene. The actual value of the gene is called an Allele; so, the eye color gene may produce brown eyes. A hierarchical picture is built up with alleles being encoded as genes with sequences of genes being chained together as chromosomes, which makes up the DNA of an individual (Whitley, 1994).

When two individuals mate, both parents pass their chromosomes onto their offspring. In humans, who have 46-paired chromosomes in total, both parents pass on 23 chromosomes each to their child. Each chromosome passed to the child is an amalgamation of two chromosomes from a parent. The two chromosomes come together and swap genetic material, and only one of the new chromosome strand is passed to the child. So, the chromosome strands undergo a crossover of genetic material, which leads to a unique new individual. As if this were not enough, genetic material can undergo mutations, resulting from imperfect crossovers or other external stimuli. Although mutation is rare, it does lead to an even greater diversification in the population. It must be noted however, that a significant number of mutations are harmful and can destroy good genetic code, so the rate of mutation must be low in order to prevent severe degradation of the genetic code.

Benefits of Genetic

Algorithm Martello (1983) as cited by Goldberg (1989) discovered that one of the GA's most important qualities is its ability to evaluate many possible solutions simultaneously.

This ability, called Implicit Parallelism (Keith Grant) is the cornerstone of the GA's power. Implicit parallelism results from simultaneous evaluation of the numerous building blocks that comprises the string. Each string may contain millions of these building blocks, and the GA assesses them all simultaneously each time it calculates the string's fitness. In effect, the algorithm selects for patterns inside the string that exhibit high worth, and passes these building blocks onto the next generation. This selection process enables genetic algorithm to perform well where traditional algorithm flounder such as in problems with huge search spaces.

The objective of this study is to use genetic algorithm (GA) to go through a number of cities that will be visited by agricultural extension agents or workers when disseminating new innovation and technology in agriculture to local farmers in order that:

- the tour is completed within minimum time (time management)
- reduction in total cost value incurred on the tour
- find the routes that minimize the total distance
- profit maximization.

Due to time constraint, the number of cities is restricted to ten (10) to give 2 x 106 initial solutions from which the fitter chromosomes can be selected for further generation. The data type of chromosomes is also restricted to only integer numbers, binary and floating point numbers are not provided for in this study

Table 1: Number of distinct solutions which exist for Agricultural Extension Agents as a function of N, the number of cities

N	1	2	3	4	6	10	20	50
No. of solutions	1	2	6	24	720	3×10^6	2×10^{18}	3×10^{64}

Source: Adewumi, 2009

Outline of the Basic Genetic Algorithm

- [Start] Generate random population of n chromosomes (suitable solutions for the problem)
- [Fitness] Evaluate the fitness f(x) of each chromosome x in the population
- [New population] Create a new population by repeating following steps until the new population is complete

1. [Selection] Select two parent chromosomes from a population according to their fitness (the better fitness, the bigger chance to be selected) 2. [Crossover] With a crossover probability cross over the parents to form a new offspring (children). If no crossover was performed, offspring is an exact copy of parents. 3. [Mutation] With a mutation probability mutate new offspring at each locus (position in chromosome). 4. [Accepting] Place new offspring in a new population

4. [Replace] Use new generated population for a further run of algorithm 5. [Test] If the end condition is satisfied, stop, and return the best solution in current population 6. [Loop] Go to step 2

Methodology

The ultimate goal of this research work is to determine a route through a set of n cities and back to starting point without repetition, which will maximize resources and save cost. N number of cities will be supplied from console for this project.

The genetic algorithm search paradigm looks for the best permutation set, which will meet the objective of the study as stated earlier. In carrying out these work, n cities representing the number of cities was permuted, which will yield n! The fitter chromosomes among them are selected for future generation. The selected chromosomes are crossed over and mutated until an optimal solution is achieved.

The Cost Matrix

The cost matrix is represented by the distances between the two adjacent cities. For example Cij represents the distance between city i and city j in cost table. The cost matrix is shown in the table below.

Table 2: The Cost Matrix

	1	2	3	4	5	....................j
1	C_{11}	C_{12}	C_{13}	C_{14}	C_{15}	....................C_{1j}
2	C_{21}	C_{22}	C_{23}	C_{24}	C_{25}	....................C_{2j}
3	C_{31}	C_{32}	C_{33}	C_{34}	C_{35}	....................C_{3j}
4	C_{41}	C_{42}	C_{43}	C_{44}	C_{45}	....................C_{4j}
5	C_{51}	C_{52}	C_{53}	C_{54}	C_{55}	....................C_{5j}
.	...	...	...	...	...	....
i	C_{i1}	C_{i2}	C_{i3}	C_{i4}	C_{i5}	....................C_{ij}

Genetic Algorithm Flowchart

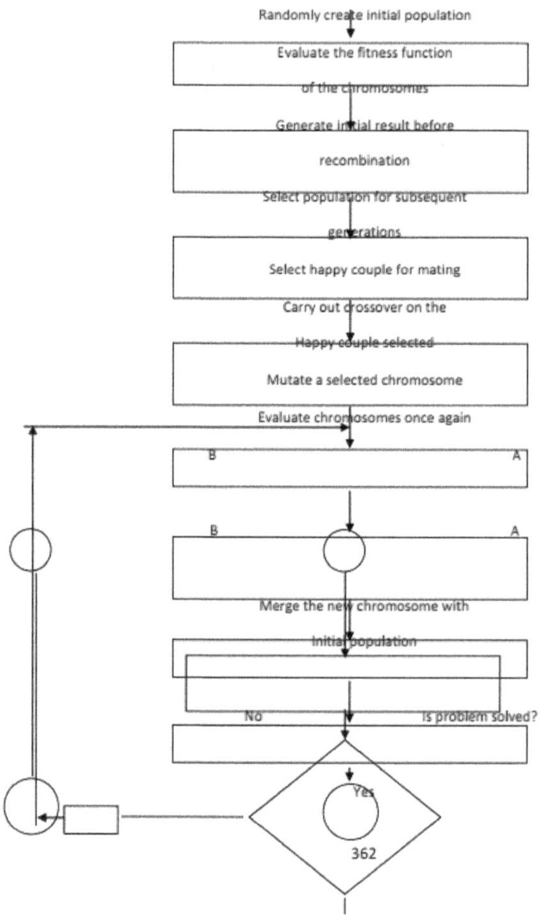

Analysis of Genetic Algorithm System for Agricultural Extension Agent

The system of transit for agricultural extension agent can be analyzed using the following components:

a. Input: The input to the system is the number of cities to visit by an agricultural extension agent. This number of cities will be captured from the keyboard or from a remote location provided that there is accessibility to the system from that location. The input data is expected to be a valid integer number. Also, a user is expected to choose stopping criteria, which will be provided. Mutation probability and crossover probability are to be supplied from the console. They should be fraction number n, where $0 < n < 1$.

b. Process: The main transformation that is expected to take place within the system is the determination of a set of solutions to solve the problems. Once a valid input data is supplied from the console, the first process that will take place is to randomly generate a set of valid tours, which will be N! where N is the number of cities supplied by the user. After the generation of all possible random tours, fitness value of each tour will be determined. Each tour in genetic algorithm paradigm is what we refer to as Chromosomes. Part of the fitter chromosomes will be selected for future mating among them. Initial population will be displayed. Thereafter reproduction operators (selection, crossover, mutation merge) will be supplied on each generation thereby leading to the production of subsequent generations. This trend continues until stopping criteria is true.

c. Output: At the initial stage, the result of initial generation will be displayed before, and thereafter, the result of subsequent generations will be displayed from where the user can select whatever option he can afford.

d. Feedback: The feedback can either be positive or negative. The feedback from this system will be whether the system is optimizing or not. If it optimizes well, the user may have to increase the value of some of these parameters or reduce the parameters depending on the parameter in question.

e. Control: This depends on the result of the feedback. If the result is favourable, the generation number may have to be increased so that more optimal solutions can be achieved. Control is demonstrated with series of input data and result obtained in appendices A and B.

f. Environment: The environment of this system that worth mentioning here is minimum hardware requirement that was used to carry out this research work. This will flourish in an environment where search space is very wide and fuzzy.

g. Boundary and interface: This system can be interfaced with other areas of artificial intelligence like fuzzy logic, simulated annealing, evolutionary algorithm.

System Development Tool

The system development tool chosen for this project is dependent on the features provided. Visual Basic Version 6.0 is used for the implementation of this design. This is due to the flexibility, simplicity of features that enhances easy movement of data from console to the program and ability to easily print results into an output file and monitor simultaneously.

System Description

This system deals with how best n cities can be toured at the lowest distance and cost possible. Visual Basic 6.0 was used to implement this and the entire program is explained below:

The main menu gives the user the opportunity to either optimize the traveling salesman problem using typical method, which will use just a method from each of the operators as follows: generational, integer representation, two-point crossover, uniform mutation, roulette selection and generation termination method. Customized optimization can also be done from this point by selecting from numbers of methods provided by each operator.

Generational Algorithms

This module offers two options if customized is selected from the main menu. The generational option randomly generates the chromosomes automatically without using any city as a starting point or terminal point. On the other hand, steady state enables you to have options of selecting the city that will be origin or destination of a chromosome.

Data Types

This customized option enables us to select what will be the format for representing cities in the chromosome. The data representations possible are binary representation, integer representation and floating-point representation. Once a data representation is chosen, it is this representation that will be used as object of manipulation for the subsequent period. The screenshot of this module is shown in Appendix B.

Crossover Operators

This module allows the user to choose the type of crossover methods to be used in case we do not want the system to choose for us. The crossover methods available in this work are partially mapped crossover, order crossover and cycle crossover. The methods are actually expected to optimize the problem, but the solution provided by these methods are relatively the same. The screenshot of the module is shown in Appendix B. the result produced by these methods are shown in Appendix C.

Mutation Operators

This module enables better result to be produced by carrying out some structural adjustment in the genes of the selected chromosome. The functionality of mutation depends on the data type selected. If the binary data type is only carried out by flipping over a bit from 0 to 1 or from 1 to 0, if the data type selected under data type module is integer representation, mutation is carried out by randomly selecting two mutation points. The genes in these two points are swapped to form other chromosomes. The mutation methods available are flip bit, boundary, uniform and Gaussian mutations.

Selection Operation

Selection operation is the reproduction operator. This operator selects chromosomes that will form population for the subsequent generation(s). the methods under this module include roulette, tournament, top percent, best and random method. Whichever method selected here will copy chromosomes from both the old and the new populations to form population for the following generation.

Termination Operator

This is the last module that determines when the genetic algorithm should stop. The methods available under this module include generation number, evolution time, fitness convergence, population convergence and gene convergence. The screenshot of this module is shown in Appendix B

Parameter specification

This module takes all the necessary parameters that will be used by the operators. These parameters among other things include the following:

Population size: This parameter specifies the number of cities that will be involved in the problem. Due to time constraints, 10 cities were considered in this work. However, this can be improved upon for it to work on several numbers of cities. Whatever number of cities supplied

will determine the number of solutions that will be provided. If population size is n, then, the population size will be n! Termination parameter: This parameter determines the number of generations that genetic operation will pass through. This will depend on the type of termination method selected; for example, in generation number, if the generation number is 10, it implies that the genetic operation will go for 10 times. Crossover probability: This probability specifies whether a crossover can be carried out on any selected chromosomes or not. The default is 0.6. The essence is to ensure that fitter chromosomes are able to survive to the following generation. Mutation probability: This probability specifies the possibility of changing the internal structure of a chromosome. It is normally kept very low so that there will be no unnecessary changes in the structure of genes. The default is 0.01.

Program Listing

The program listing of this research work is displayed in Appendix C

System Specification

The basic systems requirements to run this study are as given below:

- Windows XP Professional, Window Vista, Linux Operating System
- Pentium 233 MMX Processor
- 32MB SD RAM
- VGA Monitor

Testing

The approach adopted in testing the functionality of this program is mainly black box testing. Series of input genetic parameters are supplied from the console to get what the result will be. The operation is repeated for series of cities between 4 and 10 to see how the efficiency of the program as the search space continues to increase. Under this approach, we are only interested in the output and to what is actually happening within the program. Therefore, it may not be able to detect any redundant code that we may have within the program. In view of this, Glass Box Testing is also carried out to detect errors within the program. The methods adopted under glass box testing technique include the following: Loop testing, Branch testing, Path coverage and Statement coverage. The results generated as a result of this testing are shown in Appendix B of this work

Result Analysis

Four different parameters are supplied and their results are displayed in Appendices A and B: the first set of parameters taken so small to show the limitation of genetic algorithms over a narrower search space. Notwithstanding, the population size, if the generation number is very small with lower crossover probability and mutation probability, the GA may go through series of generation with optimizing anything at all; for example agricultural extension parameters. As the generation number continues to increase with corresponding increase in the crossover probability and decrease in mutation probability, a search space will continue to increase thereby unleashing the capability of GA approach to solving a traveling salesman problem. Figures 8, 9 and 10 in Appendix B shows the parameters supplied and their corresponding results are shown in Appendix B. However, care should be taken to vary the parameters of the transit for agricultural extension agent, the reason being that while we are trying to exploit the benefit arising from increasing mutation probability, we may also loose the initial population entirely. Therefore, we should try as much as possible to keep the probability low. The default for mutation probability is 0.01 and crossover probability is 0.6. There is no default for population and stopping criteria; it is situational, that is, it is the problem at hand that determines their values.

CONCLUSION

Genetic algorithms balance exploitation with exploration. The crossover and mutation operators control exploration while the selection and fitness functions control exploitation. Mutation increases the ability to explore new areas of the search spaces but it also disrupts the exploitation of the previous generation. It has been clearly showed from the study so far that the objectives of the work has been achieved in terms of time management, shortest routes that minimize the total distance which will leads to reduction in total cost value incurred and profit maximization for the agricultural extension agents/firm during the course of disseminating new and improved technology to local farmers in order to increase the food security in the country. The ability to separate the search mechanism from the model representation makes genetic algorithms an ideal approach for the solution of very complex combinatorial optimization problems. Travelling problem of an agricultural extension agent is one of those complex problems with a very wide search space where genetic algorithm has proved very effective to find a lasting solution to the problem of determine the least possible cost of routing a number of cities.

Recommendation

For further work on genetic algorithm implementation in agriculture, the following must be taken into consideration;

- Evaluation on a larger dataset, because in most time GA fails when it is applied to a very larger problems because it scales poorly in terms of complexity as the number of cities increases. So more research has to be carried out on the convergence of the chromosomes.

- It also has to be evaluated for a longer time frame, this is because the solution quality degrades rapidly.
- Proper fitness function has to be examined.

REFERENCES

1. Adewumi, I.O (2009). Using Genetic Algorithm For Solving Travelling Salesman Problem. B.Sc Project Submitted to Department of Industrial & Production ngineering, Faculty of Technology, University of Ibadan :2
2. Frick, A. (1998). TSPGA- An evolution program for the symmetric traveling Salesman Problem. In H.J.Zimmermann, editor, EUFIT98 – 6th European Congress on Intelligent Techniques and SoftComputing:513 – 517
3. Goldberg, D.E.(1989). Genetic Algorithms in Search, Optimization and Machine http://www.obitko.com/tutorials/genetic-algorithms/ga-basic-description.phpLearning. Addison-Wesley.
4. Martello, S. (1983). An enumerative algorithm for finding Hamiltonian circuits in a directed graph. ACMTransactions on Mathematical Software (9): 131-138. onf Odoemelam, L. E. and Onuekwusi, G. C. (2013). Effects of ICTS-Based Agricultural Extension Services in Enhancing the Well-Being of Farmers in Abia State, Nigeria. International Journal of Applied Research and Technology. 2(8): 15 – 22.
5. Oliver, I.M, Smith, D.J, Holland J.R.C (1987). A study of permutation crossover operation on the Travelling Salesman Problem. Proceeding of the 2nd International Conference on Genetic Algorithms: 224-230.
6. Prasanna J, Jung Y.S, and Dirk V.G. (1991) Parallel genetic algorithms applied to the traveling salesman problem. SIAM Journal of Optimization, 1(4):515–529 Sebusang, S. E. M. and Masupe (2003). ICT development in Botswana: connectivity for rurall communities.
7. Southern African Journal of Information and Communication (SAJIC), The Edge Institute/Research ICT Africa, Braamfortein, Z. A. (Available:http://link.wit.ac.za/journal/j0401-sebusangbotswana.pdf).
8. Torimiro, D. O. and Kolawole, O. D (2008). "New Partnership for Africa``s Development and Agricultural Technology transfer in a globalised world,`` in A. O. Akinyemiju and D. O. Torimiro (eds.) Agricultural ExtensionL A comprehensive tratise. Lagos: Ikeja ABC Agricultural systems Ltd., pp.402 – 414, (Nigeria).
9. Whitley, D.(1994). A Genetic Algorithm Tutorial. International Journal of Statistics and Computing (4): 65-85.

INDEX